The European Arc Flash Guide

A practical approach to the management of arc flash risk in electrical power systems for designers, duty holders, consultants, service providers and health & safety specialists

Mike Frain CEng FIET MCMI
European Arc Guide Ltd

Balboa Press books may be ordered through booksellers or by contacting:

Balboa Press
A Division of Hay House
1663 Liberty Drive
Bloomington, IN 47403
www.balboapress.co.uk
UK TFN: 0800 0148647 (Toll Free inside the UK)
UK Local: 02036 956325 (+44 20 3695 6325 from outside the UK)

ISBN: 978-1-9822-8406-0 (sc)
978-1-9822-8407-7 (e)

Print information available on the last page.

Balboa Press rev. date: 09/15/2021

BALBOA.PRESS
A DIVISION OF HAY HOUSE

The European Arc Flash Guide

About the Author

Mike Frain is one of the most prolific exponents of arc flash risk management and electrical safety in Europe over many years. The author is a dual national of the Republic of Ireland and United Kingdom and has been at the forefront of raising awareness about the management of the arc flash risk in a very practical way in accordance with European law. Indeed, his practical background is one of tremendous pride to him, having received a first-class electrical apprenticeship and then contributing to  electrical training and qualifications nationally and also by creating new apprenticeships. You can be sure that any advice he gives on this subject is not an academic exercise but is the fruit of his own extensive experiences. His personal mission is to inform and influence duty holders, designers, and service providers to reduce danger from electrical arcing, by providing quick, simple, accessible and accurate predictive tools coupled with practical advice.

Being an authority on the drafting and training of electrical safety procedures and processes, he has been trusted to deliver electrical safety consultancy advice for numerous household names in all sectors of industry across Europe. He has held senior management positions in electrical safety, contracting, utilities, consultancy, and facilities maintenance companies having direct responsibility for putting people to work on a full range of complex and large power electrical systems and founded Electrical Safety UK in 2004. An author of electrical safety articles and papers for several organisations such as the IET, IEEE and several engineering, utilities and HSE publications, he has been credited with raising awareness about electrical arc flash risk assessment within Europe and beyond.

The author is a Chartered Engineer, a Fellow of the Institution of Engineering & Technology and a Corporate Member of the Chartered Management Institute. He is an Expert for the British Standards Institute and IEC UK National Committee on TC 78: Live Working. He is the Convener for the International Electrotechnical Commission arc flash working group (TC 78 WG 15) and the Team Leader for the IEC arc flash end user guidance project team. He is the Secretary for the IET South Yorkshire Local Network, is Vice Chair of the IET Engineering Safety Policy Panel and leads the IET Arc Flash Working Group.

Dedication and Acknowledgements

This book would never have been published was it not for my beautiful wife Mary. For all her support throughout my career and also for her stoicism that she displayed both in sickness as she did in health. Whilst her illness meant that I would no longer be travelling the world, peering into various switch boards and small wiring panels or training engineers on electrical safety, my life was greatly enriched in many ways. Instead of her being my carer and that of our wonderful children, I became hers. I could not have been prouder than on the day that she was honoured by the Lord Mayor of Sheffield and given an award for "outstanding contribution to the community". I know that she would approve of the time that I have spent in researching and writing this book. Mary, I love you so much for the love, support, compassion, and care that you've shown to me, your family, friends, and also those countless people in need.

My grateful thanks go out to the following people for their help in by their reviews, proof reading and helpful comments.

- Claire Brady
- Michael Whitham
- Jackie Wooldridge
- Chris Ross
- Tony Pierce
- Jim Phillips
- Sean Johnson
- Dominic Parr

Table of Contents

About the Author.. V

Dedication and Acknowledgements.. vii

Introduction xv
Why did I write this book? xvii

Chapter 1   Purpose and Scope..1
1.1 What is the purpose of this guide? 1
1.2 Scope of the guide 2
1.3 Sharing Best Practice 3
1.4 Who can use the guide? 3

Chapter 2   What is Arc Flash?...5
2.1 Arc Flash Injury 5
2.2 Causes of Arc Flash..6
2.3 Consequences of an Electrical Arc Flash.............................7
2.4 Which Factors affect the Severity of an Arc Flash? 12
2.5 Blast pressure, sound attenuation & distance 13

Chapter 3   Risk Assessments and the Four P Guide 15
3.1 Hazard and Risk 16
3.2 Arc Flash Risk Model 17
3.3 Four Ps Approach to Arc Flash Management 19

Chapter 4   Prediction 25
4.1 What Does Arc Flash Hazard Prediction Tell Me? 26
4.2 Hazard Calculation Methods... 28
4.3 The Fault Level Paradox... 33

Chapter 5   Prevention & Minimisation 37
5.1 Elimination of Live Working 37
5.2 Protection Settings... 38
5.3 Improved Protection Schemes... 40
5.4 Switchgear Design.. 47
5.5 Operations Sequence 53
5.6 Remote Operation and Switching 53
5.7 Partial Discharge Testing 55

Chapter 6 Process, Policies & Procedures 59

6.1 Safe Procedures, Safe People and Safe Places 59

6.2 Safety Rules and Procedures 60

6.3 Electrical Maintenance Program 62

6.4 Signs, Labelling and Field Marking of Equipment in Europe 63

6.5 Competence 65

6.6 New Work and Alterations 70

6.7 Approval of Electrical Test Equipment & Tools 70

6.8 Thermography and Partial Discharge Testing 71

6.9 Risk Assessments – Practical Approach 72

6.10 Dynamic Risk Assessment Checklist 75

6.11 Testing for DEAD 80

6.12 Live Testing, Fault Finding and Diagnostics 86

Chapter 7 Protection 91

7.1 Standards for PPE in Europe 92

7.2 Selection of PPE 93

7.3 CE Marking 96

7.4 Types of Arc Flash PPE 96

7.5 PPE Test Methods for the Electric Arc Hazard 102

7.6 Summary of Minimum Arc Thermal Performance for Protective Clothing from IEC 61482 -2 109

7.7 Head, Face and Eyes Protection 112

7.8 Hand Protection 113

7.9 Foot Protection 114

7.10 Hearing 114

7.11 Pragmatic Approaches to PPE Selection 114

Chapter 8 Data Collection 117

8.1 Planning 118

8.2 Supply Source Data 122

8.3 The Survey 126

8.4 Protection Data 132

8.5 Low Voltage Cables and Distribution Busbars 148

8.6 Transformers 151

8.7 Motors 151

8.8 Working Distance 153

8.9 Modes of Operation 153

8.10 Consolidation of Data 154

Chapter 9 Electrical Duty Holders 157
9.1 Safe Procedures, People and Places 158
9.2 System Study 160
9.3 The Risk Gap 161
9.4 Choices of Approach to Predicting the Hazard 162
9.5 Prioritisation 165
9.6 Hazard Calculators 168

Chapter 10 Service Providers and Contractors 169
10.1 Strategy 170
10.2 The 4P Approach for Service Providers 170
10.3 What else can I do to manage the arc flash risk? 175

Chapter 11 Electrical Utilities 177
11.1 So, what makes electrical utilities special? 177
11.2 Project Overview 178
11.3 Strategy 178
11.4 Methodology 178
11.5 Modelling and Data Gathering 180
11.6 Assumptions and Generalisations 180
11.7 Results and Conclusions 182
11.8 Personal Protective Equipment 185
11.9 Protection in practice 187
11.10 Acknowledgements 187

Chapter 12 Myths & Mistakes 189
12.1 Arc Flash is much more than PPE 189
12.2 "You have to be a maths genius in order to carry out arc flash calculations!" ... 190
12.3 "An Arc Flash Evaluation is a Huge Task" 191
12.4 "We do not need to do arc flash evaluation in Europe." 192
12.5 "You will be killed in the blast above 40 cal/cm^2." 192
12.6 "Incident Energy below 1.2 cal/cm^2 means no injury." 193
12.7 Live Working on Damaged Equipment. 193
12.8 Arcing Damage to a Primary Substation 195
12.9 Hospital Design Mistake 196
12.10 Changing Busbar Tap off Units 198
12.11 Finger Safe does not equal Arc Safe 201
12.12 Competence (or lack of it!) 201
12.13 Connecting Cables into Live Equipment 202
12.14 PPE – Getting it wrong! 206
12.15 "Am I responsible for the competence of my contractors?" 207

Chapter 13 Complex Software Guide 209
13.1 Compatibility 210
13.2 Single Line Diagram 210
13.3 Fault level study 210
13.4 Protection coordination study 211
13.5 Arc flash evaluation 211
13.6 Number of buses 211
13.7 What else do you need to use the software for?...............................211
13.8 Labels..212
13.9 Technical Support 212
13.10 North American Electrical Safety Standards 212
13.11 Cost..213

Chapter 14 Hazard & Severity Calculators 215
Introduction..215
14.1 IEEE 1584 Incident Energy Calculator (LEVEL 1) 218
14.2 LV Circuit Breaker Calculator (LEVEL 2)..226
14.3 Incident Energy HRC Fuse Charts. (LEVEL 2) 233
14.4 LV Devices (Fuses and MCBs) up to 125A calculator (LEVEL 3) 235
14.5 DC Incident Energy Calculator (LEVEL 1).......................................237
14.6 Blast Pressure Calculator (LEVEL 2) 241
14.7 DGUV Box Test Algorithm Calculator (LEVEL 1)..............................242
14.8 Prospective Short Circuit Current Calculator Tool 246
14.9 LV HRC Fuse Time Current Curve Tool 253
14.10 IDMT Time Current Curve Calculator 254
14.11 Risk Assessment Forms..259

Chapter 15 Why Do Arc Flash Calculations? 261
Introduction..261
15.1 Root out danger 261
15.2 Keeping people out of restrictive PPE..261
15.3 The first step in risk assessment 262
15.4 Examples of where calculations would have prevented accidents....262
15.5 Helps to comply with the law..263
15.6 90% of the effort for a study is what we should be doing anyway ...263
15.7 Design problems are averted 263
15.8 Providing knowledge to workers..264
15.9 An essential tool for the designer 265
15.10 Improves electrical safety management..265

Chapter 16 Rules, Codes & Legislation...267
Introduction...267
16.1 Europe - Electrical Safety 268
16.2 Other EU Directives 269
16.3 UK and Ireland 270
16.4 Examples of specific legislative requirements satisfied by arc flash analysis...................272
16.5 Brexit 273
16.6 The International Electrotechnical Commission (IEC)....................................275
16.7 Directory of Useful Legislation and Standards 275
16.8 DGUV -I- 203-5188 E Guide to the Selection of Personal Protective Equipment 282
16.9 NFPA 70E Standard for Electrical Safety in the Workplace 283

Bibliography 285

Introduction

In 1989, I was summoned by the Health and Safety Executive (HSE) to attend court as a prosecution witness in a case against an electrician, which was brought under the Health and Safety at Work Act. The electrician's apprentice had been injured in an electrical flashover, or arc flash as it is often referred to nowadays and I was the first on the scene to take charge and make things safe. It was over a year later that the case came to court and I remember feeling compassion for the electrician concerned, knowing that he was a family man who had many years in the industry. At that time, he must have suffered sleepless nights as he awaited his fate. His apprentice went on to make a full recovery thank goodness. I had by that time, amassed 19 years of experience, firstly as an electrician and then as a senior manager and electrical engineer, putting people to work on complex electrical systems. I had knowledge of other electrical flashover incidents and indeed there were the tragic deaths of two electrical workers from arc flash at that same company in the late 80s. A third person was seriously injured in the incident that involved in an explosion to 11,000-volt oil filled switchgear which was being put back in service following maintenance.

In the same year that I was to give evidence in court, the Electricity at Work Regulations 1989 were published and 12 months later I was involved in writing and publishing new low voltage codes of practice for my company and training 400 contracting electricians.

My practical background had served me well in my contributions to the new safety rules and I went onto serve related working parties such as writing the National Vocational Qualifications competence standards for the electrical contracting industry and involvement with various apprenticeship schemes in addition to my day job. Indeed, my interest in the safety of electrical workers continued and 15 years later I founded Electrical Safety UK Ltd which became highly successful in writing electrical rules and supporting procedures delivering advice to large blue-chip companies in countries across Europe and in North Africa.

I have been so fortunate in the acquaintances and friendships that I have made over the years of professionals who have been passionate about the subject of electrical safety. My views on the management of electrical safety have been forged by many such people whose own practical experiences and advice have given me the incentive to write this book and pass on that knowledge.

This book is not an academic exercise or about crunching numbers through computer software, it is about identifying real hazards in the workplace and developing strategies to eliminate harm to individuals. It is with this in mind, that I have endeavoured to look globally at research and standards that could be of help. My view has been that arc flash is very simply a hazard which should be approached in the same way as any other, such as working at height for instance. If you were to put somebody to work at height you would need to know what they are doing, how high they are working and what they could fall into or onto if something went wrong. The cornerstone of European Health and Safety Directives and UK law is risk assessment, which should include an evaluation of the severity of the hazard whether it be a fall or the damaging effects of an electrical arc.

It was clear to me that there was not a great deal of research within the United Kingdom into predicting the level of harm from an electrical arc and so began my research further afield. This led me to the United States where I met my very good friend Jim Phillips who is presently the IEC international chair of the Live working committee and vice chair of the IEEE 1584 committee responsible for the guide for performing arc flash calculations. Jim and I co-authored our first article on the subject back in 2007 in the Institute of Electrical Engineer's Power Engineer magazine. It caused quite a stir at the time among professional engineers and inspired various debates and even conferences on the subject in the UK. I realised then that there was an aversion to the use of US standards in Europe as there was a perception that the first line of defence was always PPE (Personal Protective Equipment). That was never my intention, but my motivation was to use white papers and auditable standards such as the IEEE 1584 Guide for Performing Arc Flash Hazard Calculations to carry out a prediction of the level of harm. It is from that first step in the risk assessment process, that I would advocate the use of a hierarchy of risk controls always removing the hazard as the first line of defence and the use of PPE as a very last line of defence.

I am reminded of a passage from the superb book "Introduction to Health and Safety at Work" by Phil Hughes and Ed Ferrett, which is the NEBOSH National General Certificate textbook. When they address the European Council Directive 89/391/EEC (EU Workplace Health and Safety Directive), they make it clear the need to take into account such globally available information. There are nine principles of prevention specified in Schedule 1 and number 5 is "Adapting to Technical Progress". Their interpretation of this principle of prevention 5 is "it is important to take advantage of technological and technical progress, which often gives designers and employers the chance to improve both safety and working methods. With the internet and other international information sources available, a very wide knowledge going beyond what is happening in the UK or Europe, will be expected by the enforcing authorities and the courts". This guide will be based upon the European Health and Safety Directives and will be very practical in approach. It is about identifying real hazards in the workplace and developing strategies to eliminate harm to individuals. It is with this in mind that I have endeavoured to look globally at research and standards that could be of help.

Jim Phillips and I became engaged in providing guidance on how quantitative methods of arc flash risk assessment could be used to predict the level of harm from arc flash, to prevent it from causing harm and to protect individuals in the event of a flash over. Guidance was produced to comply with the law in the UK initially and eventually across Europe. Five years later, Jim and I went onto produce a paper called a European View of Arc Flash Hazards and Electrical Safety which was presented to an audience of 500 delegates at the IEEE Electrical Safety Workshop in Florida. More discussion, this time from the US side of the Atlantic and it is no coincidence that there was the adoption of a more European style risk assessment process into the US electrical consensus standards NFPA 70E Standard for Electrical Safety in the Workplace.

I knew that I was on the right track when, sharing a platform with the HSE at an IET conference in Glasgow in 2011, a question was asked by one of the delegates whether there was any likelihood that guidance on the subject of arc flash would be issued in the near future. I was pleased to hear the endorsement that if guidance was needed then I had just given it in my presentation to the conference.

Why did I write this book?

My answer to that question is: because I care passionately about electrical safety. Arc flash risk management is sometimes seen as a difficult subject with responsible persons and duty holders often confused about what they have to do to keep their workplaces safe and to comply with the law. In addition to that, there is a perception that there is great expense required in managing the hazard through time and software in particular. This is a subject area that has been very PPE centric in the last 10 years and I wanted to get the message across that preventative techniques will always be prioritised. In addition, personal injury, while being devastating for the individuals concerned is just one item on a long list of potential losses from an arc flash accident.

I have been one of the most prolific exponents of arc flash risk management and electrical safety in Europe over many years, and my desire is to share my experience and knowledge of the subject. My experience, garnered by trudging around cement mills, power stations, steel works, food companies, quarries, electrical distribution networks and countless other sectors, together with the hours of debate with foremost experts in the arc flash arena has taught me a great deal. Things like the over complication of the risk assessment process and also the common myths and mistakes that I have come across. That is why I was particularly keen to document them under a chapter of the same name, Chapter 12, Myths and Mistakes.

There is, therefore, a need to simplify the process and to make the access to the predictive tools accessible to all. Because of that, I felt that I possessed the knowledge and experience to meet this need and my personal mission is to:

> Inform and influence duty holders, designers and service providers to
> reduce danger from electrical arcing, by providing quick, simple, accessible
> and accurate predictive tools coupled with practical advice.

When I founded Electrical Safety UK Limited in 2004, one of my objectives was to be able to predict the level of harm from an arcing incident. After all, "if you can't measure it, you can't manage it". That is what first led me to the USA to learn about applying the IEEE 1584 Guide for Performing Arc Flash Hazard Calculations from first principles and using written worksheet and a calculator. Improvements in predictive techniques for the arc hazard requires designers in particular to understand that they have a duty to reduce the level of harm to people and equipment from arcing faults. Indeed, it has been my experience that more than 90% of dangerous conditions can be reduced by simple protection device alterations and adjustments. With that in mind, it is technically feasible to protect against excess current in the majority of arcing situations. Whilst this book is predominantly about arc flash from arcing faults, there is much evidence to prove that low level arcing accounts for a great deal of loss of human life and property due to fires.

Anything that I can do to shine a light on these issues and prompt more debate and future research about arcing has got to be welcomed in my view.

Chapter 1

Purpose and Scope

1.1 What is the purpose of this guide?

This guide is a practical risk assessment tool for the European Engineer,
which is primarily designed to evaluate and manage dangerous levels
of arc flash incident energy in electrical distribution systems.

It is clear from the European Council Directive 89/391/EEC (EU Workplace Health and Safety Directive) that there is an obligation on behalf of the employer to assess levels of risk involved in the workplace and the effectiveness of the precautions to be taken. For electrical work, this should include all the hazards of electricity, including the arc flash hazard, and not purely shock, as is often the case. A key element of risk assessment for electrical work is that hazards are identified and evaluated as part of a decision-making process, beginning with the justification for working on or near energised equipment.

A major objective of this guide is to provide the predictive tools to determine the arc flash hazard severity and the means to prevent it causing harm. For hazard severity, the guide provides calculators and tables to assist engineers in evaluating the magnitude of the arc flash incident energy – the first step in the risk assessment process. Whilst simple in nature, the calculators are also accurate to the extent that they are based upon the latest research. Where requirements for assessment are more complex in nature, then there is guidance on how to approach the subject of outsourcing system studies and the choices of software for modelling complex electrical engineering systems. There are several advantages to using the guide which are:

- The guide will identify significant risks quickly.
- It will provide scalability, flexibility and be a steppingstone to a more complex approach, should it be necessary.
- It is a very low cost and accessible alternative and inclusive of more members of the engineering team.
- It can easily be adopted into dynamic risk assessments.
- Easy "What if" scenarios for those times when the accuracy of input data is questionable.
- It provides a source of awareness, training & information which is in my view the biggest step change that duty holders can make.

- The guide is web based as well, so updates to standards and legislation will allow the reader to be kept up to date.
- Information is provided in developing a strategy for managing the risk.
- In depth information for carrying out complex system studies. This includes planning and pitfalls as well as information on the protective relays that may be encountered.
- There is benefit of years of shared practical experience throughout and there is a separate section on arc flash myths and mistakes.
- Use of the guide will reduce costs through:
  - Low Initial cost of the guide with associated calculator tools and accompanying tables. This is deliberately based upon the strategy of making the advice as accessible as possible.
  - Optimising PPE by taking the reader through a process of practically avoiding or reducing the risk which will often lead to more comfortable and less expensive solutions.
  - Complex software and training costs are expensive, so the use of this guide is not only highly beneficial in other ways but extremely cost effective as well.
  - Cost reduction strategy, commercial contracts, and vendor advice should a more complex approach be required.
  - Identify and group low level circuits into categories that can be excluded from complex arc flash studies.

1.2 Scope of the guide

It is important to state that the scope of the guide specifically covers the arc flash hazard and in particular the thermal effects onto a worker in case of exposure to an electric arc event, but clearly there are other electrical and non-electrical hazards associated with the task. These other hazards are outside the scope of this guide although references will be made to them where necessary and particularly helpful.

Whilst the guide is based upon European Council Directives there are some specific references to United Kingdom and Irish regulations/standards as well. Clearly, the United Kingdom left the European Union on the 31st January 2020, but their basic approach to health and safety law is unlikely to change in the near future. As the guide gives access to the tools and tables through the web version of the guide, updates to standards and legislation will allow the reader to be kept up to date. The assumption is that users of the guide will be familiar with risk assessment as defined in the European Commission document "Guidance on Risk Assessment at Work".

The guide covers the majority of situations where electrical workers may be engaged in activities where exposure to the arc flash hazard may occur. The guide is specifically applicable to working on control circuits, diagnostic testing & fault finding, live jointing, metering, LV operations, electrical installation work and maintenance activities near energised conductors and equipment. Companies who provide service engineers and electricians to client sites will find the guide to be particularly helpful in their risk assessments where intimate knowledge of the client's distribution system is not available. There is a dedicated chapter on service providers and contractors (Chapter 10) which recognises and deals with this issue in detail.

1.3 Sharing Best Practice

There is a good deal of practical experience and Chapter 12: Myths and Mysteries is based upon much of the author's personal involvement and knowledge with arc flash. The aim is to provide ongoing updates via the web version of the guide. This will cover updates to standards, regulations and calculations as mentioned previously.

1.4 Who can use the guide?

The guide is aimed at competent electrical engineers and particularly those who are responsible for putting electrical workers to task on, or in, the vicinity of live conductors and equipment. It may also be used by qualified electrical designers to supplement good design technique in order to engineer the reduction of electrical arc flash hazards and, where possible, to limit any requirement for future exposure to arc flash hazards. The calculator tools to determine incident energy within this guide are well within the capability of electrical engineers with minimum electrical design experience. For electrical networks that involve more complex feeder arrangements, the guide provides information on the procurement of software and/or service providers. The individuals named on the following non exhaustive list may find this guide useful.

- Electrical Duty Holders
- Electrical Designers
- Electrical Contractors
- Electrical Safety Consultants
- Health and Safety Managers
- Electrical Engineering Managers
- Electrical Maintenance Engineers and Technicians
- Distribution Company Employees and Service Providers
- Electrical Engineering Students

Chapter 2

What is Arc Flash?

An arc flash is usually caused by inadvertent contact between an energised conductor such as a busbar or wire with another conductor or an earthed surface. When this occurs, the resulting short circuit current can melt the conductors and produce strong magnetic fields that blow the conducting objects apart. The resultant fault current ionises the air and creates a conducting plasma fireball with arc temperatures that can reach upwards of 20,000 degrees centigrade at its centre (four times the surface temperature of the sun), which will vaporize all known materials close to the arc immediately and the incident energy emitted may ignite the worker's clothing and/or cause body burns to the worker even without igniting the clothing. Severe injury and even death can occur, not only to persons working on the electrical equipment, but also to people located nearby.

2.1 Arc Flash Injury

Arc flash injury can include external burns (i.e., severe burns to the skin), internal burns and intoxication from inhaling hot gases and vaporised metal, hearing damage, eye damage and blindness from the ultraviolet light of the flash, as well as many other devastating injuries. Depending upon the severity of the arc flash, an explosive force known as an arc blast may also occur. This is due to the rapid expansion of air, dispelling a force that may exceed 100 kilopascal (kPa) and could cause the propulsion of molten metal, equipment parts and other debris at speeds of up to 300 metres per second (m/s).

The severity of the thermal effect of an arc flash is defined by the amount of "incident energy" that a victim, standing at a given distance away from the arc, could receive to the skin surface. The "incident energy" is the value calculated which defines the severity of the arc flash. It can be quantified in units of kilojoule/metre^2 (kJ/m^2), Joule/centimetre^2 (J/cm^2) and calories/centimetre^2 (cal/cm^2). One cal/cm^2 is equal to 4.184 J/cm^2 and is equal to 41.84 kJ/m^2. Units of cal/cm^2 will be used throughout this guide, since this is the unit, which is specified to be put on the PPE garment labels according to IEC 61482-2.

As a frame of reference for incident energy, an exposure to heat flux of 5.0J/cm2 (equivalent to 1.2 cal/cm^2) during a 1 second exposure can produce the onset of second degree burn to the skin. This value is used by many standards as the benchmark that defines protection against the thermal effects of arc flash and the threshold of a zone which is commonly known as the arc flash boundary. This is where the predicted incident energy is calculated to be 5.0J/cm^2 (1.2 cal/cm^2).

2.2 Causes of Arc Flash

The following is a simplified description that I have used many times to explain the phenomena to delegates at seminars and in magazine articles. I find that the ubiquity of the components highlighted strikes a chord to give some practical understanding of the preceding description of the phenomena. As stated earlier, an arc flash is usually caused by inadvertent contact between an energised conductor such as a busbar or wire with another conductor or an earthed surface. Put simply, it is insulation breakdown and very often, the insulation in question on low voltage systems is simply air. Sadly, the cause of insulation breakdown is, very often, human intervention.

In Figure 2.1 you can see two copper busbars that are separated by 3 cm air gap. The busbars are connected to an electrical source operating at 400 V, with a prospective short circuit capacity of 25,000 A. This is roughly the type of output that you would get from a 1000 kVA, transformer. As you can see there is no protective device in circuit, only a switch. Connecting the two busbars is a piece of heavy gauge fuse wire.

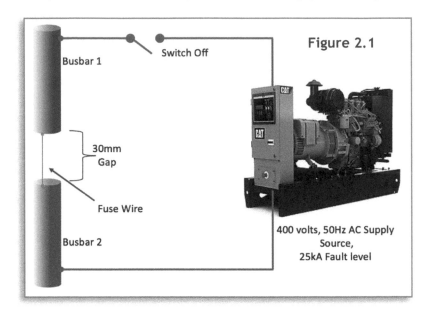

Figure 2.1

When I ask electrical workers the question "What will happen when the switch is closed," they often say that a large amount of current will flow, and the fuse wire will simply rupture and disconnect the circuit.

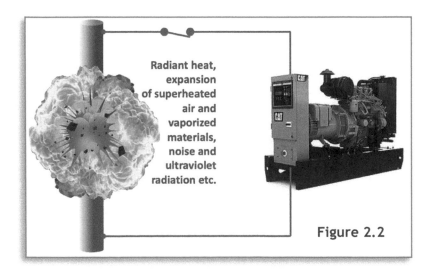

Figure 2.2

What actually happens is that the fuse wire will vaporise and in so doing so will ionise the air between the two conductors creating a low impedance path, see Figure 2.2. Large amounts of current will flow creating temperatures as high as 20,000°C at the arc which will immediately vaporise any element known to man as described previously. There will be a massive expansion of air and vaporised conductor material which causes an electrical explosion.

For instance, copper will expand up to 67,000 times its original volume in a split second. The thermal effects can cause horrific burns, ultraviolet radiation which can cause arc eye and the ballistic effects can cause blunt force trauma to anyone standing close by.

The consequences of an arc flash to such a person are clearly personal injury or even death but this can also result in fines and compensation claims to companies and individuals. A damaged brand name may also be an outcome, for instance an electrical contractor being prosecuted or served notices for a matter of core competence. If the equipment is so severely damaged that it is put out of action, then this will usually counter the reason for working live on equipment in the first place, which is to keep production flowing.

In a practical situation, the distance between busbars or conductors in low voltage switchgear can be half that stated in the example given. It is easy to imagine that the arc can be initiated not by a piece of fuse wire but by dropped tools or materials or by a short circuit in an electrical test instrument.

Once we have initiated an electrical arc, how can we extinguish it? Well, either the source of energy will collapse, or there will be such damage that an electrical arc cannot be sustained or more commonly a protective device will be used which will cut off the current, preferably to limit the damage and effect of the arc. Here is where designers have a problem. We design over current protective devices at low voltage based upon available prospective short circuit current into a zero-impedance fault. Whilst the current in an arc at high voltage is almost equal to the prospective short circuit current, the three-phase fault current at low voltage is likely to be much lower than the prospective value. The arcing fault may in fact present itself as an overload rather than a dangerous short circuit which will result in a long disconnection time.

So, what is the significance of this fact? The energy released in the arc is directly proportional to the amount of current flowing and the amount of time that the current will flow. So, if you double the disconnection time, you will double the energy released and double the amount of damage sustained. For Low Voltage circuit breakers, the difference between the device sensing a short circuit and an overload can be a factor of hundreds which will in turn create hundreds of times more energy released in the arc. More about this later.

2.3 Consequences of an Electrical Arc Flash

There is evidence to show that the majority of direct arc flash incidents occur due to and with individuals in the vicinity of the arc. My own research has revealed that the majority of fatalities due to electricity are in fact from shock but that the majority of injuries in industry are due to electrical burns. I speak about incidents that have led to loss of life in this guide that I am personally aware of and also of incidents causing injury and those where no injury has been incurred. It has been my experience, that whether individuals have been injured or not, arcing incidents cause huge losses in many instances. I have categorised some of those losses as follows.

2.3.1 Injury or death

You do not need to inquire much further than a quick internet search to reveal masses of material including photographs and videos depicting horrendous arc flash injuries to individuals. Put it simply, electrical burns are horrible. They are very slow healing and leave lasting physical and mental scars.

2.3.2 Classification of Burns

Burns are classified differently and as the subject of a second-degree burn is used in arc flash severity evaluation, it is worth briefly describing the terms used for the sake of clarity. The mildest classification is the first degree burn also called a superficial epidermal burn. The epidermis is the only skin which is affected and there may be reddening and slight swelling but no blisters. The next is a second degree burn or a partial thickness burn which affects the second layer of skin or dermis. The symptoms are as the first degree burn but blistering may occur either immediately or sometime later. Next in severity is a third degree burn also called a full thickness burn which goes through all layers of skin to subcutaneous tissue (or fat) below. A fourth-degree burn goes through all layers of tissue and affects muscles and bones. The more severe burns often result in nerve damage which means that the patient may not feel pain, but risks of life-threatening infections are a common feature depending upon how much of the body is affected.

The inhalation of toxic and superheated products of combustion are an often-overlooked consequence of arc flash. I am aware of one case where a young man was affected in this way and the damage to his lungs was so severe that he would never work again. Blindness due to ultraviolet light, thermal burns and blast is another outcome. The effects of ultraviolet light can cause cataracts to appear years later, the injuries that the apprentice forementioned in the introduction suffered temporary blindness.

2.3.3 Age and Burns

I became fascinated by the correlation between electrical accidents, age and the survivability of electrical burns. You would think that with age would come wisdom and experience but for electrical accidents that does not follow. A colleague of mine used to roll his eyes when somebody would say to him that they had 25 years' experience in the industry. He used to retort "Is that 25 years' experience or is it 1 years' experience repeated 25 times?" The following graph (Figure 2.3) shows the relationship between age and fatal electrical based accidents in the UK over a 12-year period up to 2008. What stands out for me is the fact that the highest fatal accident rate is between ages 40 to 54. But it does not stop there because the number of fatalities above the age of 60 has got to be out of step with the age demographic in the electrical industry in that period.

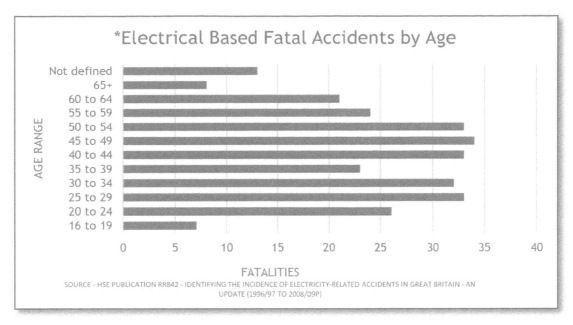

Figure 2.3

*Note - Electrical based includes all fatalities due to direct contact with electricity as well as accidents reported under other electricity-related RIDDOR categories (e.g., industries such as electricity production, installation of wiring/fitting, manufacturing of electrical appliances; occupations such as electrical engineers, electroplaters, etc.).

So, then we come back to the issue of age and the survivability of burns. An arc flash injury to seniors such as myself is definitely bad news. Even for those in their 50s the chances of survival have diminished fourfold from 60% down to 14% for 75% burns. There is a system called the Baux Score which looks at both factors of the percentage burn to the body and the person's age to forecast mortality due to burns. It simply adds the percentage of the body surface burned to the patient's age to create a score which is a prognosis indicator and a score over 140 would not be considered survivable.

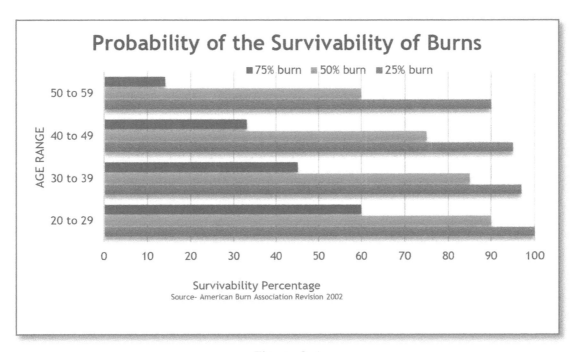

Figure 2.4

2.3.4 Fines and Compensation Claims

Whilst the financial penalties due to fines are not always as severe and eye watering as they should be, they are becoming more noteworthy. Just recently, an electrical contractor was fined £1,000,000 plus costs for breaches of section 2(1) of the Health and Safety at Work Act 1974. This was following burns to an electrician who was undertaking routine testing. In mitigation, the contractor said that a full review of electrical safety processes and systems across its sites had since been carried out, with "actions being taken to minimise the risk of recurrence". The costs of arc flash accidents do not stop there however, and compensation claims can run into many hundreds of thousands of pounds depending upon severity.

In the UK, there does not even have to be personal injury to provoke legal action by the authorities. This comes out of the requirement to report an arc flash event under the UK Reporting of Injuries, Diseases and Dangerous Occurrences Regulations (RIDDOR). This places a duty to report certain serious workplace accidents, occupational diseases and specified dangerous occurrences (near misses). To give an example, a formal report will need to be made of any explosion or fire caused by an electrical short circuit or overload which either: results in the stoppage of the plant involved for more than 24 hours; or causes a significant risk of death. There is a good chance of subsequent investigation and theoretically, the Health and Safety Executive (HSE) may wish to serve legal notices or even prosecute.

2.3.5 Damaged Brand Name

When I was working with a certain large electrical and mechanical contractor, it became very clear to me that a recent arc flash incident could potentially threaten the viability of their company. They were very well established and had a large blue-chip client base in the UK and beyond. Unfortunately, they were being investigated by the HSE following an arc flash incident involving an electrician who had been connecting a cable into a supermarket main switch panel out of hours. I was engaged by them to help them to put in place policies and procedures which should have been in in force in the first place to prevent anything like that happening. It was a bit like shutting the stable door after the horse had bolted. Senior management spent countless hours in investigations, discussions and legal representations to mitigate against the threats that a prosecution under the Health and Safety at Work Act would mean for them. From my experience as a former electrical contractor and also as a client I realised that once a prosecution had been secured, the company would have to declare that fact for every pre-qualification that they would have to complete, either for new works or for ongoing maintenance of existing contracts. When I worked as a client, I would find it extremely off-putting to see a pre-qualification from a prospective electrical contractor which declared a failure to demonstrate adherence to their own core competence criteria. The pre-qualification process is very much centred around the contractor being able to demonstrate that core competence. My message to contractors is to be mindful of the damage to your brand that could occur when working on energised equipment.

2.3.6 Severe equipment damage

I give several examples in this guide of damage to electrical equipment which has required complete replacement due to the effects of arcing. Replacement of electrical switchgear carries a high price tag not only in cost but procurement time and outages. Repairs are not often possible due to lack of replacement parts. I did some research some time ago on the age of high voltage electrical equipment in service in the UK in industrial/commercial sites. Over 40% of equipment was over 20 years old and 20% over 30 years old so there was likely to be a high level of obsolescence. I was staggered to find that there were switchgear and transformers still in service from before World War 2. (Sample size 1200 industrial and commercial sites in an estimated total of 26000 sites)

2.3.7 Lost Production

Very often the reason cited for working on live equipment is that production is so important that it would not be possible to shut the equipment down to work on it safely. I am not saying that continuity of production is not a valid reason in every circumstance, but I would say that it is not a valid reason in most circumstances. Having a shutdown in a planned way will stop production for the duration of the work whereas having an electrical flashover in switchgear will result in an unplanned event. In most cases, it is not going to be possible to resume production in a reasonable amount of time if the damage is severe because of the points made in my previous paragraph. It is for the business concerned to determine the risk through a proper risk assessment and there are the tools here to provide a measure of severity should things go wrong as part of that assessment.

2.3.8 Fires

In recent years, the subject of arcing appears to have been concentrated on the protection of an individual standing in front of energised equipment. As a result, the debate has been polarised around personal protective equipment. What we sometimes forget is that arcing is responsible for huge losses due to electrical fires and I was staggered by the statistics. In England alone in the year 2018/19 there were 4199 fires in non-domestic premises where the source of ignition was named as electrical distribution and a further 1826 fires attributable to other electrical appliances. The majority of non-domestic fires that were specified as having "non-human factors" were in fact from electrical distribution causes. I would estimate that most electrical fires are the result of electrical arcing. (Source UK Home Office)

2.3.9 Environmental Impact

As indicated previously, the justification for carrying out live work is often influenced by the essential nature of the equipment which is being serviced. For instance, a 24-hour production facility which will require some planning in order to de-energise the electrical system. What I have established is the severe consequences of an arc flash event. The same mindset of preserving power at all costs could be applied to processes that rely on power to protect the environment. There are many obvious examples in the water industry such as treatment plants and storm sewage pumping stations or the nuclear industry, but also right across industry wherever power is required to control potentially environmentally damaging processes. In my Chapter 12: Myths and Mistakes, I give the example of an arc flash event whereby a serious situation was narrowly averted when polychlorinated biphenyls (PCBs) could have been discharged into a water course following an electrical flashover in a high voltage capacitor. As always, the safer option of a planned outage is preferred to an unplanned event following an expedient quick fix repair under live conditions.

2.4 Which Factors affect the Severity of an Arc Flash?

An understanding of factors which will affect the severity of an arc flash is the key to the management of the phenomena. The distance from the arc, the arc duration and arcing current all have an effect on the severity of the arc flash in terms of incident energy but also determine the severity of ultraviolet radiation, sound pressure and blast pressure. The following describes, in simple terms, the relationship between each one of these factors.

2.4.1 Time and Current

Incident energy is directly proportional to arcing current and also the duration of the arc. In other words, if we double the arcing current then the incident energy will also double providing the arc duration stays the same. If we halve arcing current, then the incident energy will also reduce by a half. Exactly the same thing happens with the duration of the arc and by reducing the time to trip by a half will halve the incident energy. Whilst it is not that easy to attenuate the arcing current in most cases, the arc duration can be limited by faster disconnection of protective devices. This is the one very commonplace answer to reducing incident energy and therefore severity.

It may be worth mentioning here that higher fault levels may lead to faster disconnection of a protective device which can lead to a reduction of incident energy. This phenomenon, which I first identified and have referred to for many years as the "fault level paradox" will be explored in more detail in later chapters.

2.4.2 Arc in an Enclosure

When an arc occurs within an enclosure, the emission from the arc is focused outwards towards the operator. There is more likelihood that the directed effect of heat, but also the effect of hot metal particles and splashes that accompany actual electric arc faults may cause greater injury than an arc in open air. Furthermore, the size of the enclosure has an effect as well. Calculations of incident energy take into account not only whether the arc is in an enclosure or in open air, but also the dimensions of an enclosure.

2.4.3 Conductor geometry

Current studies and standards have also taken into account the configuration of the electrodes. The intensity of an electrical arc and the resultant level of incident energy is affected by the conductor geometry or electrode configuration. For instance, horizontal versus vertical conductors, open air versus enclosed and also how the conductors terminate. As we explore the topic of predicting incident energy, there will be more detail about how to apply electrode configuration in a practical setting.

2.5 Blast pressure, sound attenuation & distance

As mentioned previously, there could also be an explosive force called arc blast which can be responsible for blunt force injuries. (There are a few myths around the subject of arc blast which will be explored in more detail later) Increased distance has the effect of attenuation of the blast pressure wave and according to theoretical formulae developed by Ralph Lee in 1987 the pressure is inversely proportional to the distance to the power of 0.9. In other words, to double the distance will roughly halve the blast pressure.

According to the same formulae, the blast pressure is directly proportional to the arcing current. Therefore, a doubling of the arc current will result in a doubling of blast pressure.

Although there is much debate about how Lee's theoretical formulae can be interpreted in the real world, I have first-hand knowledge of where the ballistic effects of arc flash have caused tremendous damage and potential for injury. In addition, the fact that doors are closed on equipment will most certainly affect the likelihood of an arc flash event but will not guarantee reduced injury as closed doors will not contain the blast unless the equipment is arc protected. Arc protected equipment is discussed in Chapter 5: Prevention.

The accompanying sound pressure wave can cause permanent deafness through the rupturing of eardrums. It is sometimes incorrectly assumed that damaging sound pressure decreases as an inverse square of distance in the same way as sound intensity. In fact, it approximates to being inversely proportional meaning that a doubling of distance will result in a halving of sound pressure.

The intensity of any illumination declines as the inverse of the square of the distance. Twice as far 1/4 as strong three times as far 1/9 as strong etc. If there is smoke or dust suspended in the air it would be even faster, but these are things you would not want to count on.

Learning Points

➢ It is very easy for an arc to be initiated when working on energised equipment.
➢ It cannot be taken for granted that the electrical protection will limit the damage caused.
➢ The consequences of arcing incidents are far greater than personal injury.

Chapter 3

Risk Assessments and the Four P Guide

In my introduction to this guide, I explained my view that arc flash is, very simply a hazard which should be approached in the same way as any other. The arc flash hazard needs to be determined by risk assessment out of which, the decision to work live or dead and the required precautions will be derived. Other electrical hazards such as electric shock obviously need to be assessed but are not the focus of this guide.

A risk assessment must be performed where there could be danger which may include performing an arc flash calculation study to define the severity of the hazard. It is a fact that many workers are put to work on very high incident energy equipment without a thought for the consequences which results in a huge risk to the company and the individual. There is a need to evaluate the arc flash risk and the methods for controlling that risk as part of the process. The amount of energy that could be released in an electrical flashover is the starting point in that evaluation. Once that is known then risk control measures can be explored to determine the test of reasonableness in working live and also suitable precautions that will be required to prevent injury.

In 1989, what would be known as the "European Framework Directive" number 89/391/EEC (Workplace Health and Safety Directive) was passed which introduced measures to encourage improvements to the safety and health of workers. A cornerstone of this directive is risk assessment which means that any employer must evaluate all the risks to the safety and health of workers. For electrical safety, this means that the arc flash hazard cannot be simply ignored even though there are no prescriptive standards in place, such as in the US and Canada and certainly no direct link from the hazard directly to arc flash PPE.

Directive 89/391/EEC creates an obligation on behalf of the employer to assess the level of risk involved in the workplace and the effectiveness of the precautions in place. For electrical work, all hazards should be considered, including the arc flash hazard and not purely shock, as is often the case with many European companies. Arc flash risk assessment for workers who operate in proximity to, or on, energized electrical equipment, cables and overhead lines, is an essential part of electrical safety management. Electrical work must be carried out with conductors de-energized and isolated wherever possible and there is an extremely low tolerance for live working in many European countries.

As a result, many manufacturing plants restrict live working to inspection, diagnostic testing and commissioning purposes. There are, however, many tasks that require working either on, or near to energized equipment. Even then, it should also be acknowledged that the process of de-energization often requires exposure to the hazard through interactions such as switching, racking, and testing of equipment.

The need for risk assessment is embodied in Law through Directive 89/391/EEC and UK Management of Health and Safety at Work Regulations 1999 and the associated guidance which identifies electrical work as a "high risk" activity. As a minimum you must:

1. Identify what could cause injury or illness in your business (hazards).
2. Decide how likely it is that someone could be harmed and how seriously (the risk).
3. Take action to eliminate the hazard, or if this is not possible, then control the risk.

3.1 Hazard and Risk

Before we consider the subject further, it may be worth defining what we mean by the terms hazard and risk in the context of arc flash.

The European definition for the term hazard means anything that has the potential to cause harm. In the case of arc flash the potential to cause harm will vary with the current that can flow in an arc, the amount of time that the arcing fault is sustained, the length of the gaps between the conductive parts, which are bridged by the arc, electrodes, the confinement around the arc, the chemical compositions of the conductors and the materials around the arc, and the distance of the worker from the arc. Arc flash hazard is generally derived by system parameters.

The term risk is the chance (or likelihood), high or low, that someone might be harmed by the hazard as expressed above. Arc flash risk is generally derived from system conditions and the task being performed.

Although the arc flash hazard may be high, control measures can be adopted to reduce both the hazard, and subsequently the associated risk, to as low as possible. (Note: There is, in Great Britain and in international health and safety standards, the term as low "as is reasonably practicable" or ALARP for short. The concept of "reasonably practicable" lies at the heart of the British health and safety system and is a key part of the general duties of the Health and Safety at Work etc. Act 1974). However, many readers may not be governed by standards or laws that are influenced by such an approach and therefore, the advice that I would offer, is to ensure compliance with local and company risk tolerance when making final judgments regarding the residual degree of risk.

A fundamental safety principle is to design out, eliminate or remove the hazard at its source. This leads to the conclusion that the majority of electrical tasks must be carried out with the equipment made dead. To work dead, the electricity supply must be isolated in such a way that it cannot be reconnected or inadvertently become live again for the duration of the work. As a minimum, this will include the positive identification of all possible supply sources, the opening and locking of suitable isolation points preferably by personal padlocks and for proving dead at the point of work.

3.2 Arc Flash Risk Model

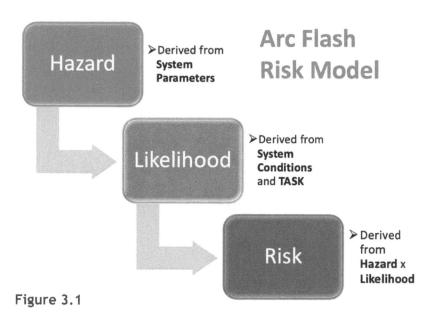

Figure 3.1

Where the arc flash hazard cannot be eliminated then suitable risk controls should be in place (preventative and/or protective measures). Figure 3.1 illustrates a simplified risk model as applied to arc flash. The hazard 'arc flash' which can be further subdivided into several hazardous effects such as thermal, ballistic, noise, and optical.

As demonstrated earlier this is expressed by the parameters of the electrical system. It is the thermal effects that are generally used to provide the potential to cause harm by way of calculations. The input to the calculations is the **system parameters** such as voltage and prospective short circuit current but equipment specific parameters as well. The protective devices such as fuses and circuit breakers are also classed as system parameters because as will be explored in later chapters, the time to clear the arcing fault is very important in regulating the severity of the hazard.

As mentioned earlier, the risk is the likelihood that this hazard may cause harm. I have therefore found it useful to highlight the likelihood separately in the arc flash risk model.

There was a time when some standards advocated the estimation of the hazard severity and providing personal protective equipment to the level of severity, regardless of likelihood. This would lead to the wearing of heavy-duty PPE to carry out normal routine operations on electrical equipment which was designed and built to an extremely high standard. What was missing were the **system conditions** and also the nature of the **task** to be carried out.

The condition of the equipment, the quality of the installation, how well it has been maintained and whether it is being operated in accordance with its original design are all factors to do with the **system conditions** and therefore will influence the likelihood. Furthermore, the physical **task** to be carried out on or near energised equipment is a hugely significant factor as it is very often worker activities that initiate a damaging arc flash event. An example of where the task can influence the likelihood is that of excavations or civil works around live cables. I remember a case which was being investigated by the UK Health and Safety Executive in the UK. It involved work within a substation and a new cable was to be drawn into a concrete duct alongside some existing live high voltage cables. Where the cables passed from one room to another, the duct had been made up with cement as a fire and moisture break. The workers proceeded to use a jack hammer to break out the cement and succeeded in penetrating the sheath, armouring and insulation of one of the live cables. The resultant fireball caused severe injury.

Having asked the question about what constitutes live work on many training courses, it is very rare that an example of working around live armoured cables would ever be mentioned. The specific references to live work in the UK (and Irish) regulations highlights work on or near a live conductor (or live part) other than one suitably covered with insulating material so as to prevent danger. Clearly the cables in my example are insulated but not suitably to withstand a power tool such as a jack hammer.

As stated above, the activities associated with the physical task can often lead to the initiation of an arc flash event. For low voltage work the following list describes some activities that have the potential to initiate an arc and some of which have been shown to be common causes of electrical flashover. Whilst some of the following activities are common in industry, practices such as connecting cables into live equipment would be very difficult to justify.

1. Connecting cables into live equipment
2. Testing; especially with substandard instruments and test methods
3. Testing on damaged cables and equipment. There are several known cases of arc flash due to using voltage indicators on faulted cables
4. Inspections or any interactions which involves the exposure of live low voltage conductors
5. Work on or adjacent to live low voltage conductors that are insulated but where the work may adversely affect the integrity of that insulation. Some examples would be drilling into panels and drawing cables into cable management systems. My earlier example about civil works around live cables also falls into this category
6. Custom and practice activities such as Installing or repairing equipment which is adjacent to exposed live low voltage conductors
7. Removal and replacement/insertion of live components such as circuit breakers in panel boards and large power bus bar tap off units
8. Live underground cable jointing
9. Live overhead line work
10. Switching and racking out poorly maintained or legacy LV switchgear
11. Replacement of fuses and links especially onto faults

We have a legal duty to carry out risk assessments in the workplace. (Note that EN 50110-1 Operation of electrical installations also states, "Before carrying out any operation on an electrical installation, an assessment of the electrical risks shall be made") We are guided that the risk assessments have to include the following steps.

1. The identification of hazards
2. The identification of those at potential risk from those hazards
3. An estimation of the risk involved
4. Considering if the risk can be eliminated; and if not;
5. Making a judgement on whether further measures to prevent or reduce the risk need to be introduced.

The first step is to define the hazards. Electrical flashover is a known hazard, which will need to be taken into account when dealing with interactions with high energy electrical systems.

There are means available within this guide to evaluate the risks in line with point 3 above, which will include an assessment of the severity of the arc flash. Note that the steps are about elimination, prevention and reduction of risk and deciding on precautions does not mean a direct link into personal protective equipment. Only when prevention techniques have been exhausted should PPE be considered.

Arc flash risk assessment for workers who operate in proximity to, or on, energised electrical equipment, cables, and overhead lines, is an essential part of electrical safety management. Electrical work should be carried out with conductors dead and isolated wherever possible but there are tasks that require working either on, or in close proximity to, energised equipment. Even then it should also be acknowledged that the process of de-energisation often requires exposure to the hazard through interactions such as switching, racking and testing of equipment.

As founder of Electrical Safety UK, I was the lead consultant for The DuPont(™) Arc-Guide which was written by European experts. This would take the user through a step-by-step approach to the management of the arc flash hazard and includes the application of the latest developments in prevention and mitigation measures. This will always start with a dead working policy as a matter of principle and then through a range of risk control measures before considering Personal Protective Equipment (PPE) as a last resort to protect individuals should an arc flash occur. I have refined the model, which is shown in the following diagram.

3.3 Four Ps Approach to Arc Flash Management

This cycle matrix diagram illustrates how the important first step of **Predict** is used to calculate the severity of the arc hazard. This is followed by **Prevent** in that we apply the principles of prevention and order the risk control measures in a hierarchy. The next step is **Process, policies and procedures** where we apply the building blocks of safe procedures, safe places and safe people. The final step is **Protect** which looks at providing personal protective equipment as a last resort which, if the previous three steps have been

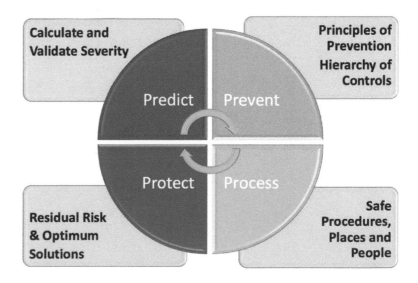

Figure 3.2 4P Risk Management Cycle

correctly applied, will cover residual risk and hopefully, result in more lightweight optimum solutions.

3.3.1 Predict

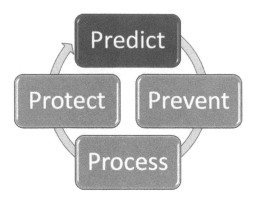

The first step in the 4P approach is to **Predict** the severity of an arc flash. In predicting the severity of the thermal effect of an arc flash, I recommend that the calculation methods are taken from the IEEE 1584 Guide for Performing Arc Flash Hazard Calculations 2018 which is an auditable standard and widely accepted in the electrical engineering community. Also covered and provided with this guide are other calculation means such as the commonly named German Box Test Method and the Ralph Lee Theoretical equations for voltages over 15kV.

The IEEE 1584 calculations take into account distance to worker, conductor gap and configuration, enclosure type voltage, prospective short circuit current and disconnection time. The output relates to the amount of "incident energy" that a victim, standing at a given distance away from the arc, could receive to the skin surface. There are also accurate calculators provided with this guide to determine prospective short circuit current, which is always a key element in predicting arcing current and therefore the incident energy levels. There are calculators and tables that will help even when site data is limited such as for circuit breakers and common European style fuses.

3.3.2 Prevent

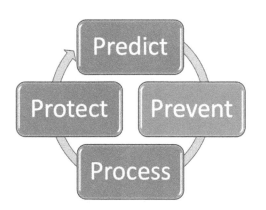

Once we have estimated the hazard level through the first step of prediction, we are guided to **prevent** the hazard from causing harm. Chapter 5: Prevention is dedicated to the fundamental principle of prevention starting with the elimination of live working but then going on to describe various practical solutions to help the reader to understand and to embrace methods and technologies that are available to reduce that harm to as low as is reasonably practical. Where we cannot eliminate the live working, then application of reduction or minimisation techniques will require a recalculation under the predict step.

3.3.2.1 Principles of Prevention for Electrical Flashover

Where the arc flash hazard cannot be eliminated then suitable risk controls should be in place with preventative measures taking the priority. Article 6(2) of European Council Directive 89/391/EEC EU (Workplace Health and Safety Directive) states "Where an employer implements any preventative measures, he shall do so on the basis of the principles" shown below. This is also specified in Management of Health and Safety at Work Regulations 1999 schedule 1. My interpretation of each of the nine principles of prevention when applied to the arc flash hazard is shown in italics.

1. Avoiding the Risk *– which means Dead working, Not energised = No electrical danger*
2. Evaluation of the risks which cannot be avoided *– by arc flash assessment and predicting the level of harm and likelihood*
3. Combating the Risks at Source *– by designing out the arc flash hazard or reducing it to an acceptable level, even as a temporary measure for the period of work*
4. Adapting to the individual *–limiting exposure to the hazard*
5. Adapting to Technical Progress/Information *– take advantage of technological and technical progress to improve both safety and working methods. The evaluation of the hazard has progressed, as have mitigation and protection techniques in respect of arc flash*
6. Replacing the dangerous with the non-dangerous *– replace vulnerable legacy switchgear and control panels preferably with arc protected equipment and/or high levels of insulation and segregation of control and power circuits. Using safer equipment (e.g. test equipment) and tools (e.g. insulated)*
7. Developing a coherent overall prevention policy *– create a **safe procedures** approach which is specific to structure environment, workforce and equipment issues and developing risk-based investment to reduce exposure to the hazard*
8. Giving collective protective measures priority over individual protective measures *– create a **safe place** of work approach by screening live parts and by good design. Any measure that is not dependent upon the individual's choice*
9. Giving appropriate instruction to employees *– create a **safe people** approach by documenting safe systems of work and training employees in safe work practices. Highlight the arc flash hazard and provide information such as in the labelling of switchgear*

These general principles of prevention should be considered against a hierarchy of risk controls with priority as given below. The top of the list should always take priority with PPE as a last resort.

3.3.2.2 Hierarchy of Risk Controls for Electrical Flashover

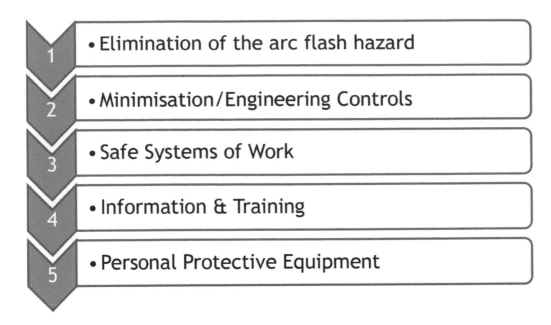

1. • Elimination of the arc flash hazard
2. • Minimisation/Engineering Controls
3. • Safe Systems of Work
4. • Information & Training
5. • Personal Protective Equipment

All these measures should be properly monitored and reviewed, and this is particularly important when considering the lower order risk controls such as personal protective equipment.

3.3.3 Process

The next step is to review policies and procedures. A review of electrical procedures and safe systems of work can substantially raise the profile and understanding of the arc flash hazard and associated control measures. Periodic awareness and refresher training, toolbox talks and specific training in the policy, rules and procedures are essential elements of arc flash risk management.

Chapter 6: Process, Policies and Procedures, and Chapter 9: Electrical Duty Holders introduce the concept of the three building blocks of electrical safety which are safe procedures, safe people and safe places. I highlighted these three facets of electrical safety in the list of principles of prevention above. (3.2.1 points seven, eight and nine)

The Process, Policies and Procedures chapter gives practical guidance on how these three facets address the specific issues of arc flash management. In particular, there is detailed information on a practical approach to risk assessment. Risk Assessments need to document the hazard severity and the risk control measures. Wherever possible these should be dynamically produced, and task based. In other words, be available to the person carrying out the work who will reassess such things as environmental conditions and equipment state. The chapter also provides guidance on competence (safe people) and maintenance (safe places). The Electrical Duty Holder chapter breaks down and explains the concept of safe procedures, safe people and safe places in much greater detail when applied to arc flash risk management.

3.3.4 Protect

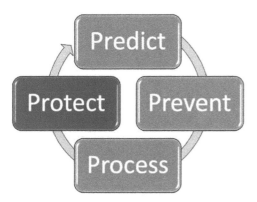

Where the risk cannot be controlled by prevention or where there is a residual risk of injury then it may be necessary to consider mitigation to prevent injury to the worker. The requirement for and suitability of mitigation techniques must form an essential element of any risk assessment. Where protection against the thermal effects becomes necessary it must be emphasised that PPE does not prevent the accident happening in the first place.

Non-flame-resistant clothing may ignite or melt at lower incident energy values and once ignited will continue to burn after the electrical arc has been extinguished. Burning material next to the flesh can result in serious third degree burns even for very short durations. This means that ordinary clothing could actually become a hazard and for this reason it should be considered within the risk assessment. Part of the risk assessment is to define the level of PPE that should be worn by the worker. However, the removal of the hazard is always the first choice.

Learning Points

➤ Risk assessments are a mandatory part of safety management which require an evaluation of risk in every case. That includes arc flash in the same way as any other hazard.

➤ Prevention is the first key principle. If the equipment is dead and correctly isolated, then there is no hazard.

➤ Never use PPE as a first line of defence against electrical flashover. It should be used as a last line of defence and any protection against residual risk using PPE is likely to be more comfortable, cheaper, and easier to justify if the 4 Ps approach is used.

Chapter 4

Prediction

The severity of the thermal effect of an arc flash is defined by the amount of incident energy that a victim, standing at a given distance away from the arc, could receive to the skin surface. In Chapter 2: What is Arc Flash, I said that an exposure to heat flux of 5.0J/cm² (equivalent of 1.2 cal/cm²) during a 1 second exposure, can produce the onset of second degree burn to the skin. This value is used by many standards as the benchmark that defines protection against the thermal effects of arc flash and the threshold of a zone which is commonly known as the arc flash boundary. This is where the predicted incident energy is calculated to be 5.0J/cm² (1.2 cal/cm²). 1.2 calories per square centimetre is often depicted as the amount of incident energy that one could receive if you were to hold your finger in the hottest part of a match or candle flame for one second. Not to be recommended but it is unlikely that you would receive anything more than mild superficial burns. Whilst

> 1.2 calories per square centimetre.

the SI unit used (International System of Units) is joules per square centimetre (J/cm²), most of the references to incident energy within this guide will be in calories per square centimetre (cal/cm²) as this unit of measure tends to be universally used for protective measures.

So, to truly understand the risks of the arc flash hazard, you need to be able to calculate the incident energy level at the worker. This first P, **Predict**, is fundamental to the whole risk assessment approach. To paraphrase Peter Drucker; "if you can't measure it, you can't manage it". This is the reason why I have put so much effort into providing the tools, calculators and tables to provide the means to predict incident energy. In using these hazard severity tools to calculate the incident energy at the point of work, this will provide the bedrock of information from which decisions can be made in respect of safely managing the risk. Once you can identify and quantify the hazard, you can decide if work can proceed and if so, what preventive or protective measures need to be in place. In addition, there are hazard severity calculators for prospective short circuit current, DC arc flash and quick tools for circuits with circuit breaker protection or fuse protection. This will allow for a figure to be obtained very quickly and easily. Once this is completed, you can continue to manage the risk by using the **Prevent**, **Process** and **Protect** chapters.

4.1 What Does Arc Flash Hazard Prediction Tell Me?

The debate about arc flash hazard management in Europe has been polarised around the need for personal protection and in the UK especially, the aversion to PPE has suppressed the true losses due to arcing faults. It is time to move on and to focus on predicting hazard severity to assist in avoiding serious injury and other consequences such as production losses, equipment/building damage, fines and compensation claims. All this before considering the provision of personal protective equipment as a last line of defence to protect against residual risk. As mentioned previously, there are tools appended to this guide which will allow you to undertake a hazard severity prediction and the following will outline what this will tell you.

4.1.1 How Big is the Bang?

Firstly, and bluntly it will tell you just how big the bang will be. You can be standing in front of two identical electrical switch panels, with one being a relatively low-level of arc flash hazard and the other one could be extremely high. In other words, one contains the equivalent of a domestic cat behind the locked panel doors whilst the other one contains a raging tiger. So, if the electrical panels are identical, what is it that causes this to be so? The answer could be very small differences in fault level or protection device characteristics or settings. The only way you will find this out is by the risk assessment process and by doing the maths. Why would you want to know?

4.1.2 Legal Requirements

To satisfy legal requirements for risk assessment which is embodied in law through European Council Directive 89/391 (EU Workplace Health and Safety Directive) and how this directive is interpreted in secondary legislation in member countries. (This is regardless of EEC past membership and further information can be found in Chapter 16: Rules, Codes and Legislation) In Chapter 3: Risk Assessment and the Four P Guide it was established that this required an identification of the hazard and then to decide how likely it is, that someone could be harmed and how seriously (the risk). Finally, we are called on to take action to eliminate the hazard, or if this is not possible, to control the risk.

4.1.3 Arc Flash Boundary

By performing hazard severity calculations, the onset of where a second-degree burn can occur, (or partial thickness burn) may be predicted at a distance from the possible arcing source. As explained earlier this figure is 1.2 cal/cm^2.

The distance at which this figure is calculated is called the arc flash boundary and is used within US and Canadian safety standards to determine the measures that will be required to protect workers. (See Chapter 16: Rules, Codes and Legislation) In Europe, this is a very useful piece of information and will allow duty holders to determine risk control measures such as a boundary within which PPE may be required as part of specific safe working procedures. Another example may be designers who need to determine working distances around electrical equipment.

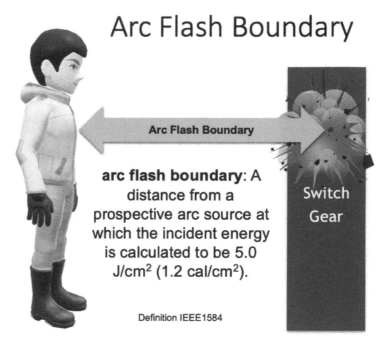

Arc Flash Boundary

arc flash boundary: A distance from a prospective arc source at which the incident energy is calculated to be 5.0 J/cm² (1.2 cal/cm²).

Definition IEEE1584

Figure 4.1

4.1.4 Fault Level Information and the Strength and Capability of Equipment

When predicting the incident arc flash hazards, knowledge of fault levels will be required. If the information is available, that is fine, but there is an online prospective short circuit current calculator to assist if not. The determination of prospective short circuit current levels is required by designers and duty holders to ensure that equipment is capable of clearing fault currents safely. This may look like a spin off from carrying out the arc flash hazard calculations, but as can be seen from the instructions, a check on the strength and capability is a necessary part of the risk assessment.

4.1.5 Competence

Technical knowledge and experience about the system to be worked on or near is a key ingredient in the assessment of competence of the individual undertaking that work. Among other things, the scope of technical knowledge or experience may include an understanding of the hazards which may arise during the work and the precautions which need to be taken. In addition, the worker must be able to recognise at all times whether it is safe for work to continue. The hazard assessment provides good information about how severe the outcome of an unplanned event will be. Although the aim must always be to avoid the hazard in the first place, there is now the technology available based on sound research to determine the protection measures for workers. There is much more information about competence in Chapter 6: Process, Policies and Procedures.

4.1.6 Personal Protective Equipment Requirements

To be able to comply with the European Council Directive 89/656/EEC (Use of Personal Protective Equipment), the selection of control measures including PPE would be very difficult to achieve without some measure of the actual hazard. (You can find the specific legislation in Chapter 16: Rules, Codes and Legislation for the UK and Ireland). PPE can only be prescribed after the employer has analysed and assessed the risks which cannot be avoided by other means. For arc flash this means that an employer must consider other means of achieving safety prior to considering the use of PPE, such as the elimination of hazard, engineering controls and safe systems of work. For further information, See Chapter 7: Protection.

4.1.7 Ballistic Effects or Arc Blast

As described within the Chapter 2: What is Arc Flash, there could also be an explosive force called arc blast which can be responsible for blunt force injuries. Whilst there are a few myths around the subject of arc blast, there are the means to calculate blast pressure in accordance with theoretical formulae developed by Ralph H Lee, IEEE Life Fellow, in 1987. According to the same formulae, the blast pressure is directly proportional to the arcing current. Therefore, a doubling of the arc current will result in a doubling of blast pressure. There needs to be some caution about the use of the formulae which is conservative at best, but this is discussed in the earlier Chapter 2: What is Arc Flash.

4.2 Hazard Calculation Methods

The accompanying calculators, tools and tables follow various published formulae and white papers but by far the most prominent is the IEEE 1584 Guide for performing Arc Flash Hazard Calculations.

4.2.1 IEEE 1584 Guide for Performing Arc Flash Hazard Calculations

The IEEE, Institute of Electrical and Electronic Engineers, is the largest professional engineering body in the world and is responsible for the IEEE 1584 Standard. IEEE 1584 is widely seen within the worldwide electrical engineering community as a valid means of determining incident energy levels and arc flash boundaries and calculations are auditable against this standard. The standard does not offer any advice or recommendations in respect of personal protective equipment or any other mitigation measures.

The IEEE 1584 calculations are based upon an empirically derived model and statistical analysis to produce mathematical equations and provides a reasonable degree of accuracy. The 2018 guide was developed from over 1860 short circuit tests performed at various voltage levels. It is important to understand however, that there are limitations to calculations based upon the standard.

The model is valid for systems having: -

- Voltages in the range of 208 V – 15000 V, three-phase.
- Frequencies of 50 or 60 Hz.
- Bolted fault current (Prospective Short Circuit Current) in the range of:
  - 700 A to 106,000 A for voltages between 208 and 600 volts.
  - 200 A to 65,000 A for voltages between 601 and 15,000 volts.
- Working Distances greater than or equal to 305mm.
- Grounding of all types and Ungrounded. (Earthed and Unearthed systems)
- Equipment enclosures of maximum height and width 1244.6mm. The width needs to be larger than four times the gap between the conductors.
- Gaps between conductors in the range of:
  - 6.35 to 76.2mm for voltages between 208 and 600 volts.
  - 19.05 to 254mm for voltages between 601 and 15,000 volts.
- Faults involving three phases only.
- AC (Alternating Current) faults only.

There are some key points to be made when interpreting the results from the arc flash calculations which are summarised as follows:

a) **Arcing Duration** has a linear effect on the incident energy explains why lower prospective short circuit current does not always correlate to low incident energy levels.

b) **Distance** from the arc has an inverse exponential effect meaning that small changes in distance can have large changes in incident energy.

c) **X/R ratio, supply frequency and electrode material.** Variations in X/R ratio of the system, the supply frequency or electrode material do not affect the results. Although the testing was at 60Hz the system of equations for calculating arcing current and incident energy should be accurate over the range of 50 Hz to 60 Hz. Conductor material was not found to be significant although other testing standards do use copper and aluminium electrodes.

d) **Arcing Current** depends primarily on available short-circuit current, bus gap (the distance between conductors at the point of fault), electrode configurations, enclosure size, and system voltage.

e) **Incident Energy** depends primarily on calculated arc current, arcing duration, and working distance. Electrode gap is a smaller factor.

f) **Earthing.** In the IEEE 2002 model there were different factors for earthed (grounded) and unearthed (ungrounded) systems. This has now changed, and the IEEE 1584-2018 model does not have system earthing configuration as an input parameter. The test results did not show any significant impact of the system grounding or bonding on the incident energy released by the arc.

g) **Arc Sustainability**. The commentary to the earlier IEEE 1584 2002 guide stated that it was very difficult to sustain arcs at lower voltages they were only able to sustain an arc once at 208 volts. The 2018 model states that "Sustainable arcs are possible but less likely in three-phase systems operating at 240 V nominal or less with an available short-circuit current less than 2000 A". At lower voltages, there is always the possibility that the arc will self-extinguish but that would be very difficult to model.

h) **Single Phase Systems**. At European single-phase harmonised voltages, (220/230 volts) we do know that severe burns have been caused to operators. However, IEEE 1584 deals with three phase faults only. Although there have been various papers that have suggested applying a correction factor to a three-phase result, I remain sceptical that this would be meaningful or accurate. A conservative approach would be to use the single-phase voltage and prospective short circuit current and then base a prediction of hazard level on a calculation of the three-phase incident energy.

i) **DC systems** are not included in the IEEE 1584 model although there has been a desire at committee level for many years to include DC testing. It is possible that DC testing will be developed in the future.

j) **Fault Types**. All testing used in the basic incident energy model was three-phase testing because three-phase arcs produce the greatest possible arc flash hazard in ac equipment. Open LV Boards and bare conductor lines where single-phase faults are likely can only be addressed as three-phase faults using the models in this guide. Other possibilities are; (i) It is widely recognised that line-to-line faults in equipment or cables often quickly escalate into three-phase faults. (ii) Low voltage system earth faults will also often escalate very quickly into three-phase faults.

k) **Arcing Current Variation Factor**. The model relies upon a calculation of the predicted arcing current value so that the operating time for the upstream protective device can be determined. I explain the fault level paradox later which talks about very large variances in incident energy due to very small changes in the operating time of protective devices due to settings or tolerances. The solution is to make two arcing current and energy calculations; one using the calculated expected arc current and one using a reduced arcing current that is lower. This is achieved by using an arcing current variation factor and then calculating a second incident energy level and arc flash boundary. The hazard severity is therefore, based upon the higher results of energy level and arc flash boundary.

The IEEE 1584 Guide for Performing Arc Flash Hazard Calculations can be purchased from the Institute of Electrical and Electronic Engineers. https://standards.ieee.org

4.2.2 German Guide DGUV-I 203-078 Arc Flash Calculations

Unlike the IEEE 1584 Guide described previously, the German guide DGUV-I 203-078 (Thermal Hazards from Electric Fault Arc) is very different in approach and the two standards cannot be compared with each other. The calculations used by DGUV-I 203-078 are derived from the box test method for PPE which is detailed in the standard IEC 61482-1-2:2014 Live Working – Protective Clothing against the thermal hazards of an electric arc. This test standard is discussed in detail in Chapter 7: Protection, but in summary, it is a simple pass/fail against just two fixed parameters of fault current, 4kA and 7kA. In other words when tested according to box test, a sample of the material that the protective clothing is made of is assigned one of two arc protection classes (APC). This will either be APC 1 which is subjected to a prospective short circuit current of 4 kA for 500ms or APC 2 at a prospective short circuit current of 7 kA for 500ms. The test is fairly simple and can be replicated on many industrial sites unlike the open arc method that requires serious decoupled motor/generator supplies.

The PPE box test method predates the calculations and more recent refinements of the mathematical models that have been achieved from work at Ilmenau Technical University in Germany. In other words, the results of the box test output have been reverse engineered to provide a quantitative method of matching site conditions to the test method. The tests to produce the IEEE 1584 calculations is discrete from the open arc tests to assess the thermal performance of PPE whereas, the DGUV-I 203-078 calculations are integral to the box test. As of 2021, the time of writing, a 2020 edition of this guide had been published in German and an English translation is expected shortly.

4.2.2.1 Merits and Drawbacks of the DGUV-I 203-078 method

Whilst the DGUV guide is used in parts of Europe for low voltage utility applications, I would recommend the use of IEEE 1584 for industrial and commercial applications. That said, the calculations are fairly simple bearing in mind that IEEE 1584 has a total of 25 formulae some of which have six variables and thirteen coefficients.

DGUV-I 203-078 considers the reactive power component for the power transferred into the arc whereas IEEE 1584 does not. It also highlights the use of arc voltage and resistance.

Unlike IEEE 1584 which is based on empirically derived data, DGUV-I 203-078 is a theoretical model. The voltage range is from 50 volts right up to 110 kV, but it is mainly applied to low voltage applications. The test rig resembles a low voltage utility service point or cut out at a voltage of 400 volts and has an aluminium and a copper electrode. The mixture of aluminium and copper outside of utility distribution termination equipment is relatively rare nowadays.

There is just one electrode configuration in DGUV whereby IEEE 1584 considers five configurations. Some may see that as an advantage but there is considerable variance of arcing current results because of the additional electrode configurations. I would trust the IEEE 1584 results as they emulate real conditions particularly circuit parameters. This is an important consideration when calculating arcing current as the two methods will produce differing results. If the arcing current is incorrect then this will have a huge impact when predicting the severity of the arc hazard.

One of the drawbacks of the DGUV-I 203-078 method is that the algorithm is reverse engineered from a test method which does not emulate field conditions in commercial and industrial setting. It is after all, a single phase 400-volt test whereas the IEEE 1584 formulae are based on real world conditions and are divorced from the PPE open arc tests. In other words, they are independent.

The following table gives a comparison between the two methods.

Comparison between the DGUV-I 203-078 and IEEE 1584 Guides	
DGUV Guide	IEEE 1584 Guide
Theoretically derived from the box test method of testing protective clothing against the thermal hazards of an electric arc to IEC 61482-1-2: Live working.	Empirically derived from actual high power laboratory tests for the full range of the guide.
Linked to the box testing of PPE.	Independent of PPE testing.
The R/X ratio required.	No requirement for R/X ratio.
Requires maximum and minimum fault currents to IEC 60909.	Calculates a minimum arcing current to account for variations in system parameters.
Note: that in all cases when obtaining fault levels, the discovery phase should account for system configuration and of utility variations as a prerequisite for using both methods.	
Arcing current theoretically derived from a single-phase low power source which cannot be matched to site conditions.	3 phase arcing current based upon 5 different electrode configurations which are based upon actual site conditions.
No allowance for enclosure dimensions.	Correction factors calculated for different height, width and depth of enclosures.
Aluminium and copper electrodes used in the box test.	No requirement for aluminium electrodes.
Calculates arc energy.	Calculates incident energy level at a working distance.
PPE restricted to standard protection APC1 or enhanced protection APC2 only.	No restriction on PPE range (except for an upper limit 100 cal/cm² -see IEC 61482-2)
Application of correction factors for transmission factor KT (enclosed, semi open and open arc) plus actual distance to the APC1 and APC2 creates an equivalent PPE protection level. This equivalent protection level must be greater than the arc energy.	PPE matched directly to the calculated incident energy.
Does not provide an arc flash protection boundary.	Calculates an arc flash protection boundary
Gives a maximum cut off disconnection time of 1 second for reaction time.	Gives a maximum cut off disconnection time of 2 seconds for reaction time.
Starts the risk assessment with PPE in mind rather than a last resort. Does not give a meaningful severity calculation.	Gives a measure of the severity of the hazard. PPE recommendations are excluded from the guide.

My preference for the IEEE 1584 method for industrial and commercial applications is given by the final point in the above table, where the DGUV method starts the risk assessment with PPE in mind rather than a last resort. The DGUV method does not give a meaningful severity calculation. To give an analogy: if you wanted to know if your workers require ear protectors, you will measure the sound levels in dB and then try to isolate or reduce if above legal limits. As a last resort you would obtain ear defenders capable of providing the necessary protection/attenuation. The DGUV does not give a severity level such as the onset of a second-degree burn. IEEE 1584 on the other hand, gives a severity level and as it says in the scope: "Recommendations for personal protective equipment (PPE) to mitigate arc-flash hazards are <u>not</u> included in this guide". In other words, you do not have to have PPE in mind when using IEEE 1584, it simply answers the question "who may be harmed and how?". PPE comes several steps later as a last resort.

4.2.3 DC Systems

DC systems are not included in the IEEE 1584 model although there has been a desire at committee level for many years to include DC testing. There have however, been several technical papers that have been published that have produced theoretical formulae for determining the incident energy levels. These are covered in Chapter 14: Hazard and Severity Calculations and the online tools provided by the European Arc Guide.

4.2.4 Warning

The prediction of the arc flash thermal hazards given above do not take into account the possibility of toxic gases, projectiles and molten metals from an arcing event. Arc faults in oil filled equipment has led to the ignition of the insulating medium leading to catastrophic explosions and fires resulting in the loss of life. Whilst these possibilities should always be borne in mind when carrying out risk assessments, it is not possible to calculate severity from the above methods.

4.3 The Fault Level Paradox

Based on the question that I asked earlier, is it the equivalent of a tiger or a domesticated cat inside the panel? The magnitude of incident energy available during an arc flash is directly dependant on the short circuit current flowing through the air gap and the time it takes an upstream protective device to clear the fault. In general, the greater the short circuit current the greater the incident energy, however this is not always the case.

It is a commonly held belief that the greater the available short circuit current is at a given location; the more damage can occur. When it comes to evaluating a protective device's interrupting and withstand capability, this is a true statement. However, in the case of arc flash, it is quite possible that a lower short circuit current can cause the upstream protective device to take longer to operate and actually increase the overall incident energy exposure.

The time current graph in Figure 4.2 can be used to illustrate this paradox. The horizontal axis of the logarithmic graph represents current in amps and the vertical axis represents time in seconds. The time current curve, also known as the tripping characteristic, defines the relationship between current and tripping time. Time current curves will typically have an inverse characteristic meaning time and current are inversely proportional to each other. The greater the current the less time it takes the device to operate, and the lower the current the longer it takes to operate.

Many protective devices will have an instantaneous trip function defined by a vertical band, as shown on the graph. If the current exceeds this value, it will trip in just a few cycles. However, if the current is less than the instantaneous value, the device will trip with some time delay.

This graph illustrates that if the short circuit current flowing to the arc is 16,000 amps, the protective device will trip instantaneously (0.06 seconds) resulting in incident energy of

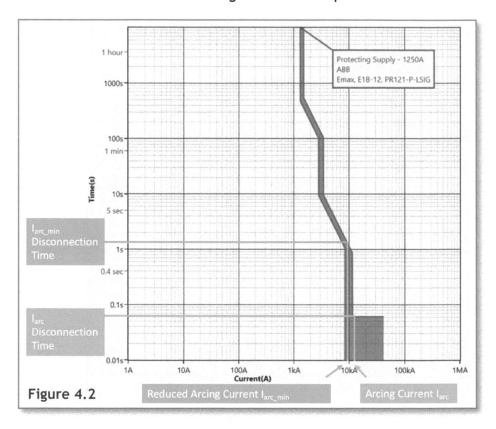

Figure 4.2

approximately 2.5 cal/cm². (Assuming IEEE 1584 VCB electrode configuration, 32mm arc gap and 450mm working distance) If the current drops to 8,000 amps, the device will no longer trip instantaneously, but instead will trip in 2 seconds. Even though the short circuit current is halved, because of the increase in the protective device's tripping time, the overall incident energy increases to over 40 cal/cm² which is seriously high in terms of thermal energy. If we were to change the electrode configuration to HCB, then the incident energy would rise to 88 cal/cm². This paradox shows how important it is that an arc flash calculation study includes various operating scenarios to evaluate the effect that the short circuit current will have on the device clearing time and ultimately on the incident energy.

I first used the term fault level paradox in several published articles over 10 years ago to describe how high fault levels could determine that the protection would operate more quickly and reduce the time that it takes to clear an arcing fault. *"Wow – who would have thought that I would have been in more danger 460 metres away from the substation than directly outside."* This was a quote from a seasoned distribution engineer when I first revealed the fault level paradox to him. We were looking at incident energy levels for live jointing operations on 400 and 500 ampere, 185mm^2 aluminium waveform low voltage feeders. The thought that the incident energy was actually quite low outside the substation was counter intuitive. It was at its highest nearly half a kilometre away.

4.3.1 What happened to good old I^{2}t?

This might seem blasphemous to a power engineer but at low voltage, fault level can often be your friend when it comes to the rapid disconnection of dangerous arcing faults. When the protection fails to clear the fault quickly it is usually because of impedance. You could say, that for low voltage, it is not the fault level that causes the danger, it is the impedance! Apart from that, the energy let through represented by I^{2}t still holds true. What I have demonstrated so far is that the incident energy depends upon both the current and the time. We must always bear in mind that other destructive consequences of arcing faults, such as ballistic effects, are definitely directly related to the fault level.

Learning Points

➢ The Prediction step is fundamental to the whole risk assessment approach. To paraphrase Peter Drucker; "if you can't measure it, you can't manage it."

➢ Arc Flash Hazard prediction is about much more than PPE.

➢ High prospective short circuit current does not equal a high thermal hazard.

Chapter 5

Prevention & Minimisation

Prevention must be the fundamental safety principle for the management of arc flash hazard. This is embodied in European legislation and cannot be over emphasised. What this means is that the Duty Holder must always seek to design out, eliminate or remove the hazard at its source.

This leads to the conclusion that most electrical tasks must be carried out with the equipment made dead. To work dead the electricity supply must be isolated in such a way that it cannot be reconnected, or inadvertently become live again, for the duration of the work. As a minimum, this will include the positive identification of all possible supply sources, the opening and locking of suitable isolation points by personal padlocks and for the proving dead at the point of work. This section is dedicated to the fundamental principle of prevention starting with the elimination of live working but then we go on to describe various practical solutions to help the reader to understand and to embrace the various methods and technologies that are available.

5.1 Elimination of Live Working

To reiterate the earlier point, electrical tasks must be carried out with the equipment made dead and isolated wherever possible. However, the actions required to make safe can expose the person undertaking the isolation. Questions of residual risk and competence of the individual will need to be considered. For instance, the safety considerations and competence of an individual making safe an underground high voltage cable to put others to work is going to be of a much higher order then that of someone isolating a piece of equipment from a local isolation point for themselves. The first example is going to require someone with a high degree of knowledge, training and experience working to strict safety rules, applying circuit main earths, padlocks, issuing permits, discharging a spiking gun and often under the supervision of a control engineer. The second example of an individual isolating for themselves carries fewer processes but the risk assessment approach is the same. I will cover this in a little more detail in Chapter 6: Process, Policies and Procedures.

Designers of electrical systems should consider the need to eliminate live work as part of the overall system design. Some of the elimination measures include the segregation of power and control circuits, safe control voltages and currents, finger safe shrouding of terminals and built-in test points.

Even then, the condition of the electrical equipment must be verified before work commences. It is not uncommon to have accidents occur on equipment that has been rendered dangerous, because electrical workers have not reinstated vital safety components such as door interlocks and insulating shielding after completion of work. I will say much more about this in the chapters on Process and also Myths & Mistakes. (Chapter 6: Process, Policies and Procedures, and Chapter 12: Myths and Mistakes)

5.2 Protection Settings

The reason why I have written a dedicated section on protection settings rather than a generic approach to protection is born out of my own practical experience in the field of arc flash prevention. Whilst protection systems will not eliminate the arc flash hazard entirely, they can prevent events from causing serious injury and damage. Having undertaken studies in industry, commerce, utilities, power stations and distribution companies, I can boldly state that over 90% of seriously damaging events can be prevented by performing alterations to protection settings.

Since incident energy is a function of short circuit current and the protective device clearing time, a reduction of the arc flash hazard may be achievable by evaluating the protective device sizes, settings and time current curves. Many times, where the incident energy is at dangerous levels, it is because the upstream protective device's instantaneous adjustment is set too high, and the device is operating in the long-time delay region. Lowering the instantaneous setting may allow faster trip times resulting in lower overall incident energy. Caution should be exercised however, because lowering a device's trip setting can create a reliability issue by compromising selective coordination with other devices. This could cause multiple devices to trip during a short circuit and lead to a more widespread outage.

The problem can be solved by changing the device settings only when live work is going to be performed. After the work is completed, the original settings can be restored to maintain existing selective coordination. These types of temporary setting changes are often referred to as "maintenance settings". There are manufacturers who have devised maintenance setting schemes for this purpose and this is described later in more detail.

To consider the point further I am going to use the example of a low voltage circuit protected by a simple 400 ampere moulded case circuit breaker with an adjustable instantaneous pick up of between 5 and 10 times the circuit breaker rating. That is, the instantaneous trip can be set at between 2000 amps and 4000 amps. Illustrated in Figure 5.1 is a time current characteristic curve for the device which has an opening and a clearing characteristic.

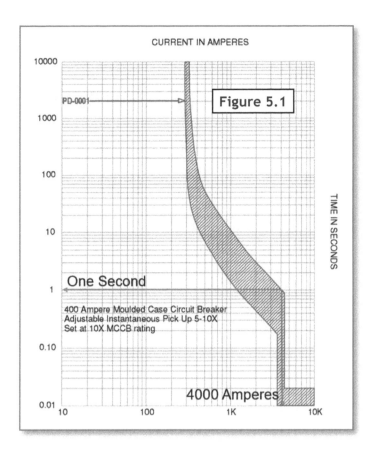

CURRENT IN AMPERES

Figure 5.1

TIME IN SECONDS

PD-0001

One Second

400 Ampere Moulded Case Circuit Breaker
Adjustable Instantaneous Pick Up 5-10X
Set at 10X MCCB rating

4000 Amperes

We are interested in the clearing characteristic for the purposes of this example which is the line to the right-hand side of the curves. If we predict the arcing current to be 4000 amps and the instantaneous setting of the circuit breaker is at 10 times (4000amps), the time for trip will be of the order of 1 second. Should an arc flash occur and be sustained for this period of time a large amount of thermal energy will be expended in the arc resulting in damage and injury depending upon the design of the equipment. Based upon arc flash calculations for enclosed low voltage equipment this could be in the order of around 11 cal/cm^2. A solution, should protection coordination permit, will be to lower the instantaneous setting to 6 times the circuit breaker rating or 2400 amps.

As you can see from figure 5.2, the resulting time to trip will be in the order of 0.02 seconds or one cycle at 50 Hertz. We have previously discussed the relationship between time and incident energy and this reduction in thermal energy is directly proportional to the reduction in tripping time. So, this simple measure, which can be accomplished by a screwdriver in this case has reduced the incident energy by a factor of 50 (1/.02) to 0.22 cal/cm^2.

My experience is that little effort has been needed to reduce dangerous fault conditions in electrical equipment because of this very simple approach. I remember working on a large technology manufacturing site which had extensive high specification clean room areas. There were 120 panels within the clean room zones that had incident energy levels above 1.2 calories per square centimetre. Working with the Duty Holder, we were able to reduce the incident energy levels below this figure by the adoption of protection alterations. The majority were comparable to the example given above. The other techniques used were to change fuse types or ratings or to reroute the source of supply from a different distribution busbar to reduce the source impedance. Quite

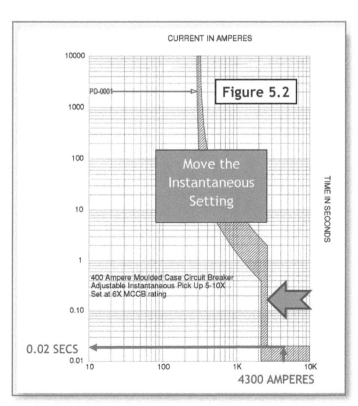

CURRENT IN AMPERES

PD-0001

Figure 5.2

Move the
Instantaneous
Setting

TIME IN SECONDS

400 Ampere Moulded Case Circuit Breaker
Adjustable Instantaneous Pick Up 5-10X
Set at 6X MCCB rating

0.02 SECS

4300 AMPERES

simple measures that were inexpensive resulted in an easy to achieve hazard reduction programme for that particular client. Modifying protection settings in order to reduce the risk is what I would term the low hanging fruit of arc flash hazard management.

5.3 Improved Protection Schemes

Protection arrangements should be explored at design stage to prevent or minimise the effects of electrical flashover using schemes such as:

- Fast acting and/or current limiting protection devices
- Maintenance Setting Schemes
- High Impedance Bus Differential
- Low Impedance Bus Differential
- Arc Flash Detection

5.3.1 Current Limiting Protection Devices

HRC Fuses. High rupturing capacity (HRC) fuses fall into the category of current limiting protection devices. In Europe the NH DIN (Deutsches Institut für Normung or German Institute for Standardization) fuses and the British Standard BS88 fuse are the most popular types of HRC fuses, and both are compliant with IEC 60269-1 standard for low voltage power fuses and both types are current limiting. The fuses are not physically interchangeable, but the characteristic curves are similar. Both NH DIN and BS88 fuses have been around for many years and have a proven reliability record. Indeed, they have remained popular with specifying engineers to this day because of their high fault current handling and minimal energy let-through (I^2t). Indeed "British style BS88 fuses are typically found in equipment manufactured in the United Kingdom or British Commonwealth countries. However, North American manufacturers have begun to specify British style fuses particularly in UPS applications at 240V or less to take advantage of their size, performance, and cost benefits". (Source Cooper Bussmann Fuses)

HRC fuses can minimise the amount of energy let through at high prospective short circuit current and therefore minimise the incident energy. See Figure 5.3. Current limiting fuses can be relied upon to operate in accordance with their operating characteristic. Circuit breakers may not if poorly maintained.

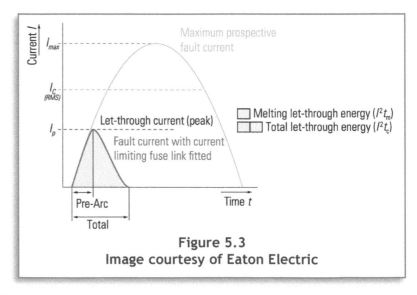

Figure 5.3
Image courtesy of Eaton Electric

A common misconception is that if fuses are current limiting, then the resultant incident energy from their use must always be lower than other overcurrent protection devices. This will only be true if the arcing current is seen by the fuse as short circuit current. As we have seen in the previous chapter, the fuse may see the arcing current as an overload and in which case the time to operate will take longer. The graph in Figure 5.4 shows the indicative incident energy level for a low voltage system which is protected by an 800 ampere BS88 high rupturing capacity fuse.

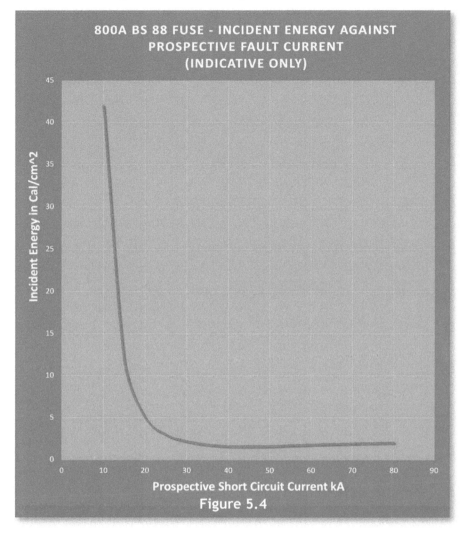

Figure 5.4

As can be seen, the incident energy at a prospective short circuit current of 20 kA or greater is below 5 cal/cm². At 80 kA the incident energy is only 2 cal/cm². Below 20 kA however, the incident energy level is much higher as the fuse will take longer to operate and as we have seen previously, incident energy is directly proportional to the disconnection time.

Circuit Breakers. There are also current limiting circuit breakers widely available for lower ratings (up to 630 amperes). To qualify as current limiting, they must be capable of restricting the let-through I^2t during a fault to no more than the I^2t available during a half-cycle of prospective symmetrical short-circuit current. This can be shown graphically in figure 5.5.

Current limiting circuit breakers are identified from the manufacturer's catalogue data of current limiting characteristics for each device. If current limiting characteristics are not available, then the device is usually a non-current limiting circuit breaker. Non-current limiting circuit breakers discharge the full short-circuit current in longer time than a half cycle and their contacts must be capable of withstanding the short-circuit current for a specified time.

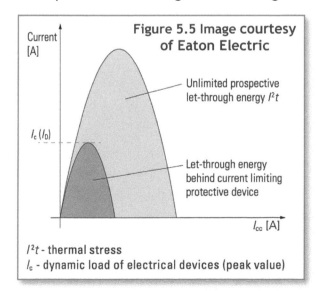

Figure 5.5 Image **courtesy of Eaton Electric**

I^2t - thermal stress
I_c - dynamic load of electrical devices (peak value)

Fuses versus circuit breakers. This is one of those debates that exercises the minds of electrical design engineers. Experience shows that preferences are often not based upon sound principles. When it comes to arc flash, there are examples in this guide of where the use of each type of protection may have been a contributory factor in an electrical flashover accident. The first example is in my introduction. I spoke about my experience as a witness for the prosecution against an electrician who had failed to isolate an electrical system prior to his apprentice carrying out work which resulted in him being injured in an arc flash incident. The point of isolation was a 200-ampere moulded case circuit breaker which had opened but the red phase contacts were welded closed due to the earlier fault condition. Therefore, it is a misconception that circuit breakers will always faithfully open all poles of supply whatever type of fault condition. Clearly, the accident could have been avoided by implementing a proper isolation process.

The second example of a contributory factor to an arc flash accident concerned with protection arrangements is apparent in the example that I write about in Chapter 12: Myths and Mistakes (Live Working on Damaged Equipment) In that case it was attributable to the use of fuses. Whilst the fuse carrier and mono block isolator were damaged by an electrician due to negligence, it may serve to highlight that changing large fuses does require a degree of skill and care.

Having opened the debate about fuses versus circuit breakers the following are commonly understood advantages and disadvantages. Fuses are relatively low cost, simple and very reliable in respect of operation, maintenance free and can interrupt high short circuit current with minimal noise or combustion products. In addition, current limiting fuses are available in sizes greater than current limiting circuit breakers. On the other hand, circuit breakers take less time to restore power, where they can be reset following an overload condition and are generally easier to discriminate between upstream and downstream devices. In respect of arc flash protection, circuit breakers give flexibility in that they may have auxiliary devices connected such as arc flash detection and have maintenance settings applied. However, fuses can be relied upon to operate beneath their time current characteristic curve for arcing short circuits whereas poorly maintained circuit breakers may not. Protection arrangements are always a compromise between overload and short circuits, the equipment being protected and coordination requirements. As a result, both fuses and circuit breakers will commonly be used in the same electrical distribution system.

5.3.2 Maintenance Setting Schemes

Interactions with energised equipment of any description will inevitably lead to some risk. These may be inspections, testing, maintenance, operations or perhaps more intrusive tasks that have been properly assessed and stand the test of reasonableness. A solution may be to retrofit circuit breakers with instantaneous trip units or to temporarily alter the instantaneous settings on circuit breakers. Equipment manufacturers are continually developing better methods to reduce or eliminate the arc flash hazard and have come up with their own solutions to incorporate maintenance settings to the circuit breakers. One such system is described below.

Eaton Electric recently patented a system that they call "Arc flash Reduction Maintenance System(TM)" (ARMS technology) for low voltage moulded and air circuit breakers. It allows for a maintenance mode to be set on the circuit breaker which will in effect provide a lower instantaneous pick-up level via a maintenance switch, which is illuminated when activated. When fitted, it operates independently of the main protection unit and is designed to trip with no intentional delay once the instantaneous threshold has been exceeded.

Arc flash Reduction Maintenance System (ARMS)

The following example, Figure 5.6 shows how they ARMS system is deployed and as can be seen in this example this will reduce the time to trip from 250 milliseconds down to 25 milliseconds. As discussed previously, the incident energy is directly proportional to time so in this case the energy reduction is 90%.

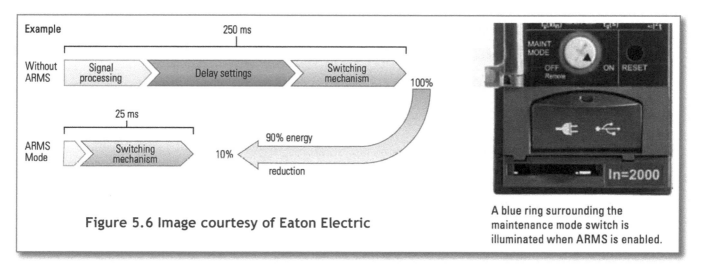

A blue ring surrounding the maintenance mode switch is illuminated when ARMS is enabled.

Figure 5.6 Image courtesy of Eaton Electric

The following time current graph in Figure 5.7 shows the normal circuit breaker settings in blue, and the maintenance setting is in red.

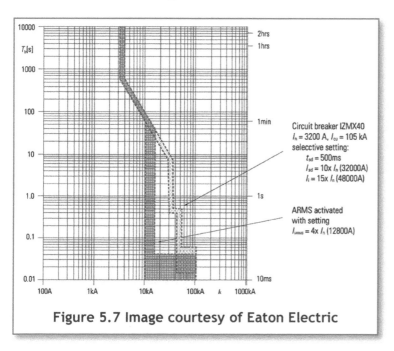

Circuit breaker IZMX40
I_n = 3200 A, I_{cu} = 105 kA
selecctive setting:
t_{sd} = 500ms
I_{sd} = 10x I_n (32000A)
I_i = 15x I_n (48000A)

ARMS activated with setting
I_{arms} = 4x I_n (12800A)

Figure 5.7 Image courtesy of Eaton Electric

As can be seen this effectively reduces the instantaneous pickup level of the circuit breaker in the same way that permanent settings are made to an existing circuit breaker as described in the previous section on protection settings. As with all modifications to protection systems, the selection of one of the reduction settings should be determined and selected by a person who is experienced in power system analysis. I can see that this type of system will grow in popularity but is a solution that is available at design stage only.

5.3.3 High and Low Impedance Bus Differential Protection Schemes

Bus Differential Protection Schemes have been used for many years, predating much of the work of arc flash calculations. They are very reliable but can be expensive. The principle of operation is based upon Kirchhoff's law which states that the vectoral sum of all currents at a node or bus is equal to zero. Looking at Figure 5.8 the equipment to be protected is the switchboard and associated incoming and outgoing circuit breakers.

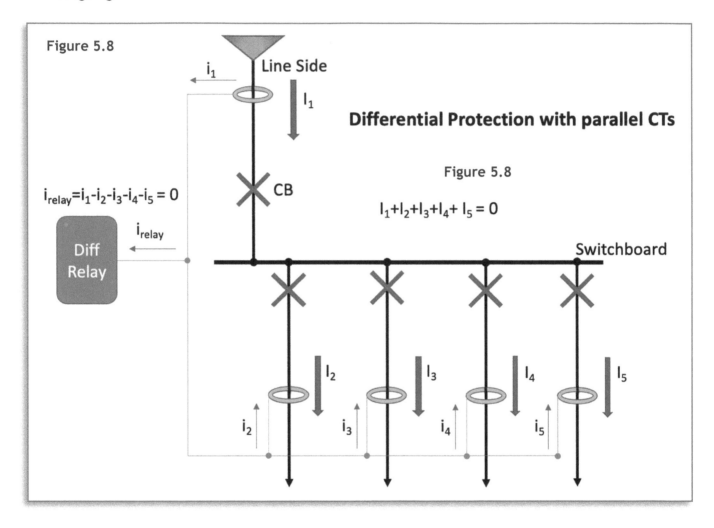

Figure 5.8

Line Side

Differential Protection with parallel CTs

Figure 5.8

$i_{relay} = i_1 - i_2 - i_3 - i_4 - i_5 = 0$

$I_1 + I_2 + I_3 + I_4 + I_5 = 0$

Switchboard

Diff Relay

CB

The current transformers (CTs) monitor the current entering and leaving the switchboard and for healthy conditions a summation of all the currents will equal zero. If there is a fault on the switchboard, or with any of the circuit breakers, then the relay will detect the differential summated current which will no longer be zero. The relay will quickly trip all of the circuit breakers both incoming and outgoing thus isolating the fault and for arcing faults this can significantly reduce the damage caused. The response time is typically 20 to 30 milliseconds and combined with a typical circuit breaker operating time of 60 milliseconds this will normally give a total clearance time of less than 100 milliseconds. Whilst cost has often put off design engineers from including bus differential schemes in the past, there appears to be a resurgence in the popularity of such schemes in preventing damage from arcing faults in recent years.

There are two types of bus differential schemes, high impedance and low impedance. As the names suggest the high impedance relay presents a high impedance to current flow whereas the low impedance relay does not. The current transformers for the high impedance scheme are simply connected in parallel with the relay. Any difference in current will create a voltage drop across the relay which in turn will trip the circuit breakers included in the scheme. Because of the sensitivity of the relay, it is critical that all the current transformers match by having the same ratios and accuracy. However, one major advantage of the high impedance scheme is scalability. In other words, if another circuit breaker is added, all that is required is a matching current transformer which is simply connected in parallel with the other CTs.

Low impedance relays, on the other hand, have low impedance to current flow from the current transformers. The difference in the installation is that each CT output has to be brought back separately to a discrete input on the relay rather than being connected in parallel with the other CTs. An advantage is that the matching of CTs is no longer critical and it is possible to share the output from CTs for other relays or meters. This makes the low impedance scheme ideal for retrofitting to existing equipment. However, it is not as scalable in that each CT will require its own relay input.

5.3.4 Arc Flash Detection

Another protection method that has been developed uses an optical sensor that detects the ultraviolet light emitted from an arc flash and causes the protective device to instantaneously trip. The method monitors the current into the equipment that is protected by the optical sensors and combined with the ultraviolet light flash from the arc and the abnormal spike in current it will quickly remove the power source and prevent serious damage. Operation time is extremely fast, usually less than 10 milliseconds plus the time for the circuit breaker to clear the fault which is typically around 60 milliseconds. The system has grown in popularity over the last 10 years, and I have seen it retrofitted to very old high voltage switchgear and also to new low voltage factory-built assemblies. The main manufacturers claim tens of thousands of installations worldwide citing the fact that they are a very cost-effective alternative to traditional busbar differential schemes on medium voltage switchgear.

5.3.5 Active Internal Arc Protection

Several manufacturers have produced innovative ways of reacting to an arcing fault by effectively creating a zero-impedance short circuit across the busbars of equipment. Sometimes the energy is dissipated to earth. Depending upon the method used for sensing the development of an arc, they can be incredibly fast and claim disconnection times as low as 4 milliseconds. They are built to withstand voltages above 40 kV and can handle prospective currents of up to 100 kA. In Figure 5.9, there is a diagram of a system that uses arc detection utilising current and optical sensing as described previously. The arcing fault is detected by the current detection unit and simultaneously by the optical sensors which then establishes a three-phase, metallic short-circuit at the earthing switch. The resulting controlled flow of earth fault current is then finally shut down by the feeder circuit-breaker. This all happens almost immediately and whilst there is the expense to consider, the resultant damage from an arcing fault is almost nullified. This means that the equipment can be put back in service quickly resulting in reduced downtime.

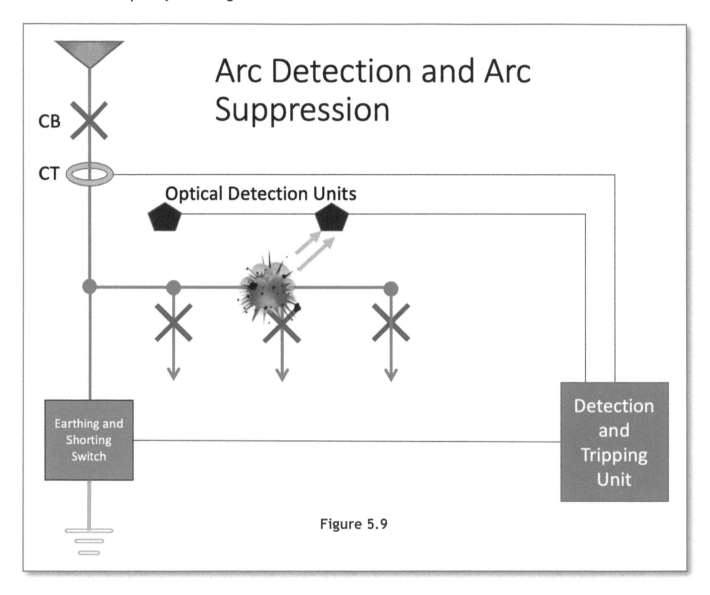

Figure 5.9

5.3.6 Arc Fault Detection Devices

Although Arc Fault Detection Devices (AFDDs), also called Arc-Fault Circuit Interrupter (AFCIs), are primarily aimed at fire safety, the common thread here is that they protect human life from the effects of arcing. The detection and clearance of arcing faults before causing injury or damage is at the heart of much of the beneficial research that has been centred around the arc hazard. Although the technology has been around for over 20 years and required by US and Canadian standards for many years, it has recently been adopted in Europe. In Germany AFDDs are required in hospitals, nurseries and high value buildings. Conventional circuit design has centred upon the clearance of zero impedance short circuits and earth fault current, neither of which will clear all dangerous low-level arcing events. Earth fault protection will pick up line to earth arcing faults but not line to line parallel faults or series faults (i.e., those involving wire breaks) For that reason, AFDDs are worth a mention here to inform of the existence of the technology and possibilities for the future in arc flash detection.

AFDDs work by continuously monitoring the wave form of the current which flows in the protected circuit. Arcing current presents itself as a high frequency and the electronics within the AFDD can differentiate between arcing fault current and those arcs that are produced by switches, contacts, thermostats, carbon brushes, etc.

Currently, the typical cost of an AFDD is in the region of £150 per device which is why they tend to be selected for high-risk situations for instance, premises with sleeping accommodation, risk of fire due to the nature of processed or stored materials, thatched or timber-framed buildings and high value locations such as museums and listed buildings. Most manufacturers provide combined arc fault, over current and earth fault detection.

They are aimed at final circuits at present, up to 63 amperes, so not an option for high incident energy situations. However, one large manufacturer told me that that current capacity will expand in the future.

5.4 Switchgear Design

Designing out the arc flash risk is high ranking on the hierarchy of risk control and above safe working systems, signs & labels and personal protective equipment. It is not however, a silver bullet which will guarantee safety on high power electrical systems. **Things change and people change things**. For example, throughout the lifetime of any piece of equipment, it may be subject to unforeseen environmental changes creating stress or damage to equipment and/or human interventions which will fundamentally undermine the way in which the original design purpose was intended for. The design, however well intentioned, needs to be a part of the safe systems, safe people and safe place's philosophy. This requires a holistic approach taking into account factors which will influence safety over the lifetime of the equipment. It cannot be over emphasised how important communication is in the provision of distribution design philosophy.

5.4.1 Holistic Approach to Design

I often hear Engineers say that they have designed out the arc flash risk but, what they are really referring to is switchgear design. Designs against arc flash require much more than that. Specific factors that need to be considered when designing out or minimising the risk of arcing are as follows.

Distribution System Philosophy

a) Review operating philosophy. For instance, can parallel or very large transformers be avoided?

b) Determine the effects of arcing when designing protective equipment.

c) Can additional protective or arc reduction devices be installed? As described in this section.

d) How is the philosophy for arc prevention communicated? For instance, through design, installation, commissioning and operating manuals.

Switchgear Design

e) How will risk assessments be implemented? Such as the identification of hazards and potential failure mechanisms.

f) Consider arc detection and suppression.

g) Consider internal arc resistance.

h) Improved protection devices set to minimise arc energy.

i) Consider remote operation.

j) Consider dual settings on protective devices.

Operation, Installation and Maintenance

k) Commissioning. The feedback loop to design from final commissioning seems so obvious but sadly often overlooked. Tales of transit packaging left inside protection relays or factory settings on circuit breakers years after installation are all too common. Commissioning processes should be informed by correct procedure and manufacturers recommendations.

l) Operation and Maintenance. What access to switch gear will maintenance teams require? What level of competence do they have and under what safety rules and processes do they operate?

m) Ensure appropriate test, inspection and preventative maintenance processes are in place. These processes should refer to arc flash precautions specifically such as relay setting checks, integrity of insulating shrouding, panels & fastenings and ingress of moisture.

n) How will the major risk of unauthorised modifications be prevented?

In Summary, a holistic approach is much more than designing out the arc flash risk through switchgear design. Minimising of the risk of arcing requires whole life care through distribution philosophy, system design, commissioning, maintenance, auditing, and operation. To achieve this requires clarity of policies through leadership and control. Neglectful dilapidation of equipment such as control panels is a tangible indicator that the policy is not working. The diagram in in Figure 5-10 shows the relationship between the Distribution Philosophy, Switchgear Design and Operations and Maintenance.

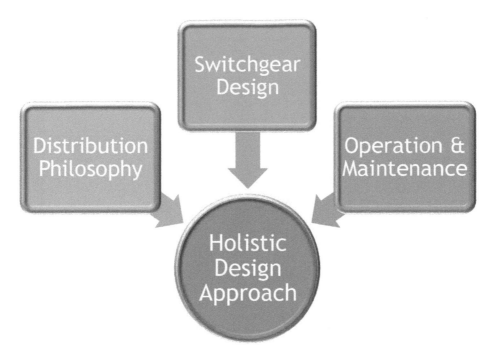

Figure 5-10 The relationship between the Distribution Philosophy, Switchgear Design and Operations and Maintenance.

5.4.2 Design of Low Voltage Switchgear

At low voltage, the design of high-power switchgear is often bespoke in nature and termed as a Power Switchgear and Control Gear Assembly. The standard IEC 61439-2 Power switchgear and control gear assemblies (PSC) applies to such assemblies and stipulates the responsibilities of the end user, the original manufacturer, and the assembly manufacturer.

The assembly manufacturer (manufacturer of the switchgear and control gear assembly) takes responsibility for the completed assembly, but this is often not the original manufacturer of the equipment. They are responsible for the manufacture, conformity to standards, including those of the original manufacturer and routine tests. The user or customer may be represented or assisted by a designer who has, in my view, the greatest responsibility in providing specific design characteristics for the assembly. They will have to provide information such as supply and load characteristics, the environmental conditions and details of operation and maintenance. Dependent upon this data, the final design of the switchgear could vary widely from one assembly to another. There is no specific international standard that covers arc containment in low voltage equipment so it is vitally important that the user (or user's designer) understands how the specification can help prevent the propagation of an internal fault. There was a time when the risk of internal faults was ignored completely and the term "fault free zone" was used to describe an area of a switchboard that had low risk. Experience shows however, that where there is a current carrying conductor, there is the possibility of a flashover. In Figure 5.11 you can see the wide difference in specifications for the insulation of busbar systems.

Figure 5.11 Non-Shrouded and Shrouded Bus Bars

IEC 61439-2 also identifies the four main categories of separation: Forms 1, 2, 3 and 4 within assemblies. I have included references to free publications that describe the forms of separation with some detail and clarity. Put simply, the higher the form number, the higher the degree of separation between terminations, functional units and busbars. So, it follows that the higher the number, then the higher the level of safety and the greater the number of tasks that can be carried out whilst the switchgear is energised.

But we cannot rely solely on the design of equipment for safe working and therefore any activities must be covered with risk assessment and safe systems of work. In arriving at which form of separation will be required, reasons for access whilst the equipment is energised will need to be explored. Some reasons may be as follows, although justification is another matter.

a) Routine maintenance, where applicable.
b) Adjusting settings of overloads and protective devices.
c) Resetting Overload or short circuit protective devices.
d) Changing fuses.
e) Taking instrument readings or measurements
f) Fault finding.
g) Re-configuring control circuits.
h) Replacing components.
i) Terminating power and/or control cables.
j) Changing the configuration of outgoing circuits.

One point of contention is how the forms of separation are interpreted for interactions with energised switchgear. For instance, there are references in some manufacturer's information that state that Form 4 construction is suitable for the connection and disconnection of cables whilst adjacent circuits are live. Whilst the source is named as the BEAMA Guide to Forms of Separation for Low Voltage Switchgear and Control Gear Assemblies to BS EN 61439-2 there is an extremely important caveat given in that document under the section on "Other Considerations Item 7." This states the following. *"Safe working with adjacent equipment energised - Switchboard manufacturers cannot give all embracing assurances for safe working, according to the form of separation with parts of the assembly energised. Specifying a particular form of separation <u>will not guarantee this for any given form number</u>. Effectively this means that <u>where live working is being contemplated, a risk assessment and judgement must be made for every situation by the Duty Holder.</u>"*

To illustrate this point, compliance with IEC 61439-2, Power switchgear and control gear assemblies (PSC) requires that the separation between functional units, separate compartments or enclosed protected spaces is attained by either one or more of the following conditions:

- Protection against contact with hazardous parts. The degree of protection shall be at least IPXXB.
- Protection against the passage of solid foreign bodies. The degree of protection shall be at least IP2X.

Both methods are similar and often referred to as "finger safe" although it is just method IPXXB that uses a test finger to prove compliance. In either case, a 12mm gap will be acceptable and this is enough to allow the passage of some tools, fastenings and cable armour wires. The main message here is to be aware that equipment design alone does not ensure safe working on energised equipment. I have referenced forms of separation in my chapter on Myths and Mistakes later and give examples where misinterpretation of design criteria has led to accidents in Form 4 equipment which is the specification for higher forms of separation.

For many of the reasons for access to low voltage power switchgear and control gear assemblies, safety can be enhanced at design stage by reducing the voltage of all controls preferably to 24 volts DC. In addition, by moving them to separate cabinets or enclosures there is a reduced requirement for working on or near energised circuits.

5.4.3 Low Voltage Switchgear and Control Gear Assemblies, Internal Arc Protection

Switchgear that is termed as arc resistant uses various methods to contain or redirect the arc away from the operator or from causing damage to the equipment or surrounding environment. Typically, such equipment will use systems such as venting or containing the arc and international standards are being refined constantly to reflect the progress in this regard.

Internal arcs within switchgear are not common but the severity can be catastrophic. Causes of internal arcs can be the result of incorrect maintenance, vermin, ingress of water or dust, dielectric faults to insulators, terminations, current and voltage transformers. Design issues such as overstressing of functional units and susceptibility to overvoltage and transients are also contributory factors. Not forgetting good old wear and tear and ageing.

I stated previously that there is no specific international standard that covers arc containment in low voltage equipment. There is, however, an IEC document, IEC/TR 61641, Enclosed low-voltage switchgear and control gear assemblies - Guide for testing under conditions of arcing due to internal fault. This is a technical report which provides guidance on testing under conditions of arcing due to arc faults (not an IEC standard) and is the document to consider. Most low voltage switchgear and control gear assemblies achieve compliance by safely exhausting the arc to safety and away from operators. This is by the use of ducts, chimneys or vents and some examples are shown in Figure 5.13.

Figure 5.13 Arc exhaust ducts.

Compliance with IEC/TR 61641 does not afford protection against injury from arcing faults whilst enclosure doors are open and as we have learnt, two thirds of arcing events happen given that scenario.

5.4.4 High Voltage Switchgear and Control Gear Assemblies - Internal Arc Protection

The international standard IEC 62271-200 High-voltage switchgear and control gear applies to equipment above 1000 volts. An option of the standard characterises the ability of the equipment to contain the effects of an internal arc fault whilst reducing the risk of burns to individuals in the vicinity. It provides a rating referred to as internal arc classification, often referred to as arc resistant classification. The concept of internal arc test has been extended to low voltage assemblies by the IEC Technical Report IEC/TR 61641 as mentioned in the previous section. Note that the internal arc type test is an optional test within IEC 62271-200 and conformity is not mandatory. Clearly, the internal arc rating has not prevented the actual arcing event but seeks to prevent injury to operators and in some cases the public who may be in the vicinity of the equipment. The rating is given for switchgear which is operating under normal conditions which means that no doors or panels are open. To achieve the rating, the equipment will be tested in a high-power laboratory where it is subject to an internal arc fault. The switchgear is surrounded by cotton indicators which simulate the presence of a person and one of the criteria to achieve compliance is that the indicators will not ignite under fault conditions.

In addition to protection to individuals from the thermal effects of an internal arc, the standard requires the verification that there are no mechanical injuries. No switchgear panel or door must open, and no dangerous parts may be ejected. The IEC 62271-200 classification concentrates on reducing risk of personal injury and does not provide any criteria for the resulting damage within the enclosure which could be extensive. Also, another effect of internal arc is pressure stress in the switchgear which must be vented away from personnel. For indoor switchgear, consideration must be given to how this pressure is dissipated to avoid damage to the switchgear building perhaps through overpressure relief systems.

To be comprehensive, the following list of criteria must be fulfilled to achieve compliance.

1. Covers and doors remain closed. Limited deformations are accepted.
2. No fragmentation of the enclosure, no projection of small parts above 60 g weight.
3. No holes in the accessible sides up to a height of 2 metres.
4. Horizontal and vertical indicators do not ignite due to the effect of hot gases.
5. The enclosure remains connected to its earthing parts.

5.5 Operations Sequence

Changing the system operating configuration (switching to an alternate feeder or engaging maintenance mode settings) may reduce the arc flash hazard. Care has to be exercised on complex networks however, as switching and paralleling supplies can result in multiple transformers feeding into a possible fault during the switching operations. This can lead to increased fault levels, even though it may be for short durations. I have known of situations where this has resulted in the over stressing of switchgear.

5.6 Remote Operation and Switching

Remote operation and switching of switchgear has been around for many years. As a concept, it is about preventing injury to the individual only, so it is not the complete answer. In other words, it does not prevent the dangerous occurrence at source and in the UK there is an obligation to report such events to the Health and Safety Executive. This obligation is expressed through the Reporting of Injuries, Diseases and Dangerous Occurrences Regulations (RIDDOR). That said, the systems are there to prevent personal injury and as we learnt earlier incident energy levels reduce rapidly with distance. As I have already mentioned, some of the systems have been around for many years and were used extensively in power companies and large industry when operating equipment that had been reported as defective elsewhere in the organisation or country. Perhaps there had been a failure detected on a particular make/model of switchgear for instance. It may or may not have been responsible for injury, but the details would have been reported on the National Equipment Defect Reporting System. It would be up to the particular power company to issue a Suspension of Operating Practice which could change the way in which the equipment would be operated. This would often mean that circuit breakers would need to be operated from a remote location or have access restrictions around certain items of plant. The National Equipment Defect Reporting System is administered by the Energy Networks Association on a subscription basis and is open to asset owners and service providers on a worldwide basis.

Various systems of remote switching are shown in the images displayed starting with Figure 5.15 which shows a lanyard system.

As you can see, this looks rather crude, made up of rope and pulleys, but it was very effective at giving good

Figure 5.15 Remote lanyard system: image courtesy of Clydesdale Ltd.

distance between the operator and in this example an 11kV oil filled fused switch. Electrically powered actuators were developed for the operation of manual switchgear which could be retrofitted

Figure 5.16 Remote circuit breaker actuator unit

if necessary. As you can see from Figure 5.16, this is fitted to an outdoor SF6 circuit breaker. The drive was to reduce downtime on electrical distribution systems but would add to safety by removing the individual from the direct operation of the switchgear.

There is a growing demand on switchgear manufacturers to provide remote operation at design stage. In Figure 5.17 you can see an example of a low voltage switch board designed and manufactured as a bespoke factory-built assembly which was destined for a European client. The earlier rope and pulleys have been replaced with a remote switching unit with an umbilical cord.

The greatest risk to operators of electrical switchgear tends to be in racking operations. This is where a removable circuit breaker is racked out or racked into the fixed portion of switchgear onto busbar spouts. The operation accounts for the deaths of two engineers that I am personally aware of at a power station where I subsequently provided an arc

Figure 5.17 Remote circuit breaker actuator unit

flash study. Most large manufacturers of switchgear are providing remote racking solutions nowadays for new switchgear. There is also demand for remote circuit breaker racking devices and remote switch actuators, operators and kits for older and even obsolete switchgear and independent companies have developed solutions over a number of years. Figure 5.18 shows a rotary remote circuit breaker racking system which is fairly universal and there are fitments that allow for the horizontal or vertical operation of circuit breaker trucks at low and medium voltages.

Figure 5.18 Remote circuit breaker racking unit. Image courtesy of CBS ArcSafe

5.7 Partial Discharge Testing

Arc flash events in medium voltage equipment are often preceded by partial discharge. It is defined as a discharge that only bridges part of the space between two conducting elements and is due to voids, cracks and inclusions within the insulation. Partial discharge in electrical switchgear and cables occurs because of ageing from thermal and mechanical stresses sometimes brought on by heat, humidity and other external influences. Left undetected, partial discharge can lead to catastrophic failure and electrical arc flashover. Prevention of a flashover due to partial discharge is therefore a worth looking at as an overall strategy.

There is a good deal of research that suggests over 80% of all failures in high voltage electrical equipment is a result of partial discharge, so early detection can prevent injury due to catastrophic events and the consequential losses of power failure. Partial discharge presents as heat at the source of the discharge, ultraviolet and visible corona, electromagnetic radiation, ultrasonic radiation and audible sound.

Figure 5.19 Image courtesy of Megger Ltd

The following methods can be used to detect partial discharge (PD) and provide condition monitoring of PD levels over time to warn of impending failure in high voltage cables, switchgear, transformers, generators and motors.

5.7.1 Electromagnetic Methods

Electromagnetic (radio) waves at around 200+ MHz are produced from the site of the discharge which induce small currents in surrounding metalwork. The high frequency voltages that are produced are known as Transient Earth Voltages (TEV) and can be picked up using a capacitive probe and handheld detector or by UHF reception ultrasonic and audio microphones.

5.7.2 Acoustic Methods

Partial discharge causes sound which is both audible and ultrasound (which is outside the audio range). This will present itself as a fizzing sound and can be picked up using sensitive microphones.

5.7.3 Thermal and Optical Methods

Partial discharge will create heat which can be picked up using thermal imaging. PD will also create ultraviolet light due to the ionisation process which can be picked up by line of sight. However, most industrial indoor enclosed high voltage equipment would not make these methods a viable solution.

5.7.4 Oil Diagnosis

Although increasing amounts of electrical equipment are moving away from using oil as an insulating/ cooling or arc quenching medium, there are large amounts out there and will be for some time to come. Oil sampling and dissolved gas analysis will, over time, monitor the condition of insulating oil in equipment and may indicate serious impending breakdown due to internal arcing and or partial discharge activity. Tests on transformers should be carried out annually in my view as results of organic acidity, dielectric strength and moisture content may indicate the conditions in which the equipment will deteriorate more quickly which may lead to partial discharge.

Although partial discharge testing has been used in low voltage systems, detection is generally restricted to equipment operating at voltages greater than 3kV.

5.7.5 Portable PD Testers

Portable partial discharge testers have come on a long way and require much less expertise and technical knowledge to operate. Whilst they will give a traffic light type output, they are very useful as part of dynamic risk assessments and also to signal where further investigation will be required. They are handheld and are capable of picking up the characteristic partial discharge noise, vibrations and RF radiation through a TEV (transient earth voltages) sensor.

Figure 5.20
Image courtesy of Megger Ltd

Learning Points

➤ Prevention must be the fundamental safety principle for the management of arc flash hazard.

➤ It is a legal requirement to design out, eliminate or remove the hazard at its source.

➤ Simple modifications protection settings can account for 90% of risk reduction techniques.

➤ New technologies in switchgear design can vastly reduce the risk and prevent injury.

➤ Reducing disconnection time and increasing distance are key risk control measures.

Chapter 6

Process, Policies & Procedures

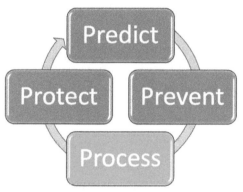

Very often electrical workers, management and duty holders do not fully understand the nature of the arc flash hazard and the seriousness of the injuries that can be sustained. As a result, experienced personnel are frequently involved in these types of accidents. Competent electrical workers should be trained in the decision-making process necessary to determine the degree and extent of the hazard and job planning necessary to perform the task safely.

A review of electrical procedures and safe systems of work can substantially raise the profile and understanding of the arc flash hazard and associated control measures. Periodic awareness and refresher training, toolbox talks and specific training in the policy, rules and procedures are essential parts of arc flash risk management.

"Most electrical accidents occur because individuals are working on or near equipment:

a) *which is thought to be dead, but which is in fact live;*
b) *which is known to be live but those involved are without adequate training or appropriate equipment, or they have not taken adequate precautions."*

This is a quotation that I have used many times and is taken from the UK HSE guidance booklet Electricity at Work – Safe Working Practices, HSG 85. This defines the key issues that need to be addressed and sets out a starting point for anyone reviewing electrical policies and procedures or maybe writing them for the first time.

6.1 Safe Procedures, Safe People and Safe Places

Chapter 9: Electrical Duty Holders expands on the concept of the three facets of electrical safety which are safe procedures, safe people, and safe places. Whilst this chapter deals with safe policies and procedures, it also overlaps with safe people in dealing with competence and guidance on maintenance procedures which impact on safe places.

Safe People
Competence, information, training & instruction. Supervision & control.

Safe Procedures
Risk assessment approach, specific rules with clear responsibilities & authorisations.

Safe Places
Well-designed & documented installations, maintenance and confidence in protection systems.

Figure 6.1 Building Blocks of Electrical Safety Management - Theme courtesy of Guardian Electric

Chapter 9, Electrical Duty Holders, breaks down and explains the concept of safe procedures, safe people and safe places in greater detail when applied to arc flash risk management.

6.2 Safety Rules and Procedures

High voltage safety rules and procedures are often straightforward to create and implement because the interactions with equipment and the competence of staff is easier to define. Low voltage rules and procedures can be more difficult because the nature of the work often entails the use of many more staff with different levels of competence. The interactions on or near live equipment is likely to be much higher and those undertaking the work may be contractors as well as staff. It is because of this complexity at low voltage, that rules and procedures need to be very specific to the organisation and environment. As a minimum, a key ingredient should include a robust set of safety rules with the following elements:

a) How the rules address company health and safety policy.

b) Clear responsibilities of stakeholders for implementation, planning, control, authorisations, audit and review. This should identify duty holders in respect of an electrical safety management system encompassing safe people, safe places and safe procedures. See Chapter 9: Electrical Duty Holders.

c) Specific authorisations of the individual for electrical activities which implies a rigorous assessment of competence.

d) Establish dead working as a principal requirement.

e) Specify controlled circumstances for live proximity work such as diagnostic testing, running adjustments and inspections.

f) Have clear instructions on where live working is strictly forbidden such as the removal and replacement of live components.

g) Auditing, monitoring and review.

To supplement the rules there can be separate procedures which can be amended or added to without the need to rewrite the electrical safety rules. They can cover specific equipment on site such as maintenance, working on generators or chiller plant for instance, but the following are some examples of underpinning procedures for the safety rules.

a) Permit to work system for work to be undertaken on the electrical system.

b) Authorisation and assessment process for any electrical work by staff and contractors.

c) Approval of electrical test equipment, tools and PPE.

d) Live testing risk assessment procedure.

e) Audit process for switchgear and control panels.

f) Signs and labels for switchgear.

g) Emergency procedures and first aid.

h) Information, instruction, and training.

i) Minor works and alterations.

There should also be local guidance written into the rules and procedures to help with interpretation and avoid ambiguity. The following flowchart Figure 6.2 shows the relationship between company policy, rules, procedures, and guidance.

Figure 6.2

Safety rules and procedures with clear responsibilities are essential parts of a safe system of work. They should deal with properly assessing and authorising only competent people and should have systems for risk assessment. Where live working cannot be avoided, then the safe working systems should stipulate the use of the correct equipment and instruments. Electrical flashover accidents are very often caused by the operator dropping uninsulated tools or metal parts or by using incorrectly specified instruments. There should be a rule that no live work will be allowed on equipment that is damaged even for the reasons of proving dead. There have been many incidents involving damaged cables where an approach has been made to prove dead when the damaged cable was in fact still live.

6.3 Electrical Maintenance Program

The main objective of an electrical maintenance program is to keep electrical equipment in good working condition so that it can reliably and safely operate within its design criteria. This includes testing and maintenance of circuit breakers, protective relaying and associated equipment on a regular basis. Improper maintenance of equipment can contribute to the severity of an arc flash. When a protective device such as a circuit breaker or relay is not properly maintained, the likelihood of it operating slower (or not at all) increases, which would also increase the duration of exposure to the arc flash. Although not technically part of an arc flash calculation study, a proper equipment maintenance program is vital in making sure the protective devices will operate correctly.

Maintenance inspections in particular can substantially impact on the control of arc flash risk with minimal additional effort in most cases. Visual inspections by competent individuals can pick up many defects that raise their own awareness of the hazard and reduce the likelihood of future accidents. In addition, good intelligence on vital assets can inform maintenance and replacement strategies.

In the case of control panels in industry, individuals interact with the equipment on a regular basis for production and breakdown purposes. Sometimes, it is these individuals who are responsible for the accumulation of defects that could lead to fatal electrical shocks and arc flash incidents. Issues such as forgetting to replace vital shrouds or fastenings or possibly allowing unauthorised temporary lash ups to become permanent can have fatal consequences for their colleagues at a later date. It is not just the individuals who are working inside the panels that create safety critical defects. I have worked extensively as an electrical safety consultant in the food industry with many household brands as well as presenting to the Institute of Occupational Safety and Health on the subject. Hygiene standards and electrical safety standards in the industry can sometimes conflict with one another. The need to ensure that production equipment is kept clean means routinely employing aggressive cleaning techniques using water jets and chemicals that often leading to water ingress into electrical equipment. To quote a food industry electrician "if a jumbo jet landed in our car park and we were asked to clean it – it would never fly again."

The environmental conditions that commonly prevail in industrial production facilities means that design standards governing water/dust ingress and earthing/bonding must be of a high order. But, the first question a designer should ask is, does the electrical equipment (such as control panels) need to be in the high-risk production environment or can it be installed remotely?

All that said, duty holders must take a pragmatic view and for existing production lines it is not going to be possible to undertake any serious redesign of electrical systems. This is where routine inspections really come into their own. An audit process for switchgear and control panels, which is part of the electrical safety rules and procedures will have the effect of highlighting serious defects/omissions. This can help inform risk assessments for future interactions and at one extreme, this may be to ban certain operations until improvements or replacements have been made.

6.4 Signs, Labelling and Field Marking of Equipment in Europe

Risk Assessments must be recorded and field marking of equipment may be required which is produced following survey and inspections. Safe systems of work, safety rules and procedures as discussed previously will be required and disseminated to all affected persons. It should be noted that arc flash labelling is not a mandatory requirement unless the company concerned is working to consensus standards such as the US NFPA 70E Standard for Electrical Safety in the Workplace. The European approach is one where labelling is less prescriptive but there is a general requirement that employers must provide safety signs where there is residual risk. The regulations require employers to ensure that safety signs are provided (or are in place) and maintained in circumstances where there is a significant risk to health and safety that has not been removed or controlled by other methods. The Health and Safety (Safety Signs and Signals) Regulations 1996 is the relevant legislation in the UK and is based upon the European Directive 92/58/EEC. The legislation lays down the minimum requirements for provision and use of safety signs at work the purpose of which is to encourage the standardisation of safety signs throughout the member states of the European Union such that safety signs, wherever they are seen, have the same meaning. Signs are only appropriate where use of a sign can further reduce the risk.

Warning signs must be standardized across Europe in a way which will minimize the potential problem that may arise from linguistic and cultural differences between workers. U.S style labels based on ANSI Z535 are not acceptable in Europe as the colours and symbols do not conform to the European legislation. When designing labels, the requirements of this legislation for warning and prohibition signs will need to be in accordance with this directive. Warning signs for instance must be yellow (or amber), they must be triangular, and the yellow must take up at least 50% of the sign. Information from arc flash studies can be included such as the incident energy level, PPE and various approach boundaries. The table in Figure 6.3 gives the meaning of each colour.

Figure 6.3 European Safety Signs and Symbology (excluding Fire Signs)

MEANING OR PURPOSE	INSTRUCTION AND INFORMATION	EXAMPLES
RED Prohibition Sign or Danger Alarm	Prohibiting dangerous behaviour, STOP such as no access to unauthorised persons.	 Live Work Prohibited
YELLOW or AMBER Warning Sign	Warning of Hazards such as electric shock or arc flash.	 High Voltage
BLUE Mandatory Sign	Specific behaviour required such as wear PPE or carry out a risk assessment.	 Hearing Protection in this Area
GREEN Emergency Escape or First Aid Sign	Shows doors, exits, escape routes, first aid equipment and facilities.	

The following are typical arc flash labels that have been produced to comply with the colours and symbology used in Europe and are based upon the European Directive 92/58/EEC.

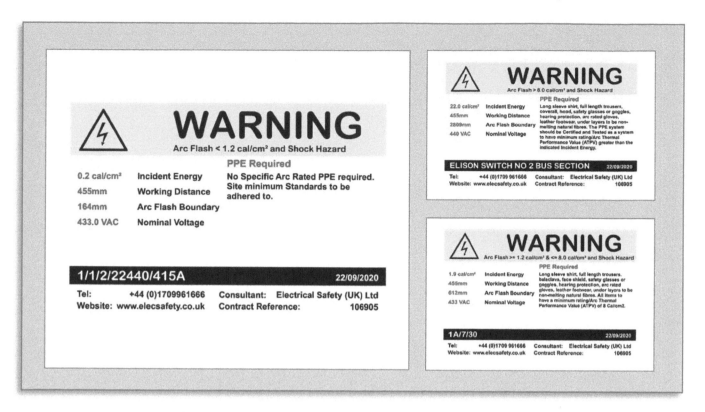

Figure 6.4 Images courtesy of Electrical Safety UK Ltd

6.5 Competence

A competent person can be described as a person who has the skill, knowledge, attitude, training, possess the experience to undertake the work and has the ability to recognise the extent of their own competence and leading them to take appropriate action. Regulation 16 of the UK Electricity at Work Regulations 1989 and the Republic of Ireland Guide to the Safety, Health and Welfare at Work (General Application) Regulations 2007 Part 3 Electricity, states that "No person shall be engaged in any work activity where technical knowledge or experience is necessary to prevent danger or, where appropriate, injury, unless he possesses such knowledge or experience, or is under such degree of supervision as may be appropriate having regard to the nature of the work". The guidance on the above goes on to say "The scope of technical knowledge or experience may include:

a) Adequate knowledge of electricity;
b) Adequate experience of electrical work;
c) Adequate understanding of the system to be worked on and practical experience of that class of system;
d) Understanding of the hazards which may arise during the work and the precautions which need to be taken;
e) Ability to recognise at all times whether it is safe for work to continue."

The very important guidance clauses c), d), and e) are greatly enhanced by risk assessments that includes the prediction of the arc flash hazard. It is not enough to say that a person is competent because he or she is a time served electrician or engineer. Indeed, my experience is that enforcing authorities expect that duty holders can demonstrate that individuals are competent for the work that they are expected to do.

6.5.1 Competence Assessment

To be authorised and deemed competent to work on or adjacent to electrical systems, employees must demonstrate the required combination of knowledge, training and experience. The rules ought to stipulate that the authorisation be carried out in writing using a certificate of competence. In that way, there are no doubts about what the individual is authorised to do as well as what they are not allowed to do.

So, where do we start in assessing competence? For existing direct employees, it is probably best to start with a framework for competence followed by an appraisal comprising an interview and a short technical competence test. Managers are often mistaken in the belief that this will be somehow demoralising; when in fact, experience shows that it can be motivating to find that one's skills are being formally recorded and recognised in such a way. Indeed, there is good evidence to show that the human brain is wired for competence. To quote the author and neuroscientist Dean Burnett, "Our brains desire a sense of competence and when we feel that we're competent, we're more likely to be happy". In 2015, the Guardian newspaper looked at several surveys to try and find the happiest occupations and concluded that the job of engineer came out on top. Being an engineer, that is no surprise to me. I have been so fortunate to have enjoyed a degree of control, competence and being fairly well paid which is often associated with the role. In conclusion, if competence assessments are followed up by formal periodic reviews and built into personal development plans, this can have a beneficial effect on morale.

6.5.2 Competence of Contractors

The competence of contractors also needs to be addressed which is exemplified in Chapter 12: Myths and Mistakes where a contractor was killed on a manufacturing plant. The company ordering the work was prosecuted for (i) not ensuring that the subcontractor was sufficiently competent to perform such work and (ii) not ensuring a safe system of work was in place. The authorisation and assessment procedure mentioned under safety rules previously, should be implemented for any electrical work by employees and contractors. I have on several occasions found engineers working on energised high-power electrical equipment with no formal electrical qualifications at all. Large refrigeration control equipment regularly contains high current busbar systems in excess of 800 amperes at low voltage.

However, electrical training is purely supplemental to some refrigeration engineers or perhaps a skill that has just been acquired over time.

For contracting companies and agencies, it is suggested that their employees must satisfy the requirements of the company safety rules and procedures in a similar way to direct employees. Temporary certificates of competence can be issued to contract employees for specific construction, maintenance or repair projects. There needs to be some recognition of emergency situations but there needs to be dialogue in anticipation of unplanned events.

6.5.3 Competence of Accompanying Persons

It is very important to consider the competence of an accompanying person when live working is involved. It is necessary to consider the role of this person in the achievement of safe working, in other words: why are they there? And what are they to do? The usual role is to be able to electrically isolate the installation, system or equipment without endangering himself or herself when something goes wrong and to render help to the worker carrying out the live work activity. They can also play a significant part in securing and managing the work area. The usual training, assessment and authorisation of such persons is necessary and appropriate criteria must be applied.

6.5.4 Competence Matrix

The creation of a competence matrix which will detail the minimum or preferred attainment in each area of competence is essential. Firstly, define a list of work categories for which the employee will receive an authorisation. For instance, the employee may be given an authorisation as a competent person carrying out dead work only. Another example is as a competent person testing and commissioning or perhaps as a competent person supervising others. Against each competence will be the minimum or preferred attainments. A list of these attainments could be as follows but clearly there could be many others depending upon the complexity of the work.

a) **Professional or craft qualifications.** Maybe these can be grouped for simplicity so one group could be Professional Qualification, Degree, relevant professional body or National Vocational Qualification Level 5 (NVQ5). The second group could have a relevant Engineering at HNC or NVQ level 4, the third a recognised apprenticeship, NVQ level 3 and so on.

b) **Formal Training in Health and Safety.** This can be carried out in house, but I can recommend the Institution of Occupational Safety and Health (IOSH) courses that can be delivered by accredited trainers on site. There is IOSH Working Safely which is a one-day course for people at any level in the organisation. For those in a supervisory or managerial capacity, I would recommend the IOSH Managing Safely which is usually four days in duration plus a day practical assessment. A need for a higher qualification may be thought appropriate so there may be another attainment level on the competence matrix for a qualification at NEBOSH

(National Examination Board in Occupational Safety and Health) National General Certificate in Occupational Health and Safety or higher.

c) **Experience**. A requirement for appropriate experience in years.

d) **Externally Assessed Competence Cards**. For electrical contracting workers in the United Kingdom the requirements of the Electro-technical Certification Scheme (ECS) is a very good start. The scheme is administered by the Joint Industry Board (JIB) for the Electrical Contracting Industry. It not only covers the traditional electrical contracting electrician but other electro-technical disciplines such as Electrical Fitters, Instruments Technicians, Maintenance Electricians, Wireman/Panel Builder, Building Controls, Emergency & Security Fitters, Cable Jointers and Telecommunications Fitters. Before a card can be issued the individual will have to prove technical and vocational qualifications, age and minimum experience and health & safety awareness. It is externally assessed and good value requiring very little administrative work on behalf of the company.

e) **Specific Vocational Training**. This classification could include formal training on particular pieces of equipment or perhaps electrical installation wiring regulations.

f) **Continuing Professional Development (CPD)**. A requirement could be to highlight ongoing enhancement of skills. This could be important where the competence of an individual may be dependent upon upskilling and keeping up to date with latest developments in a particular field or piece of equipment.

g) **Review and Audit Requirements**. There may be a formal review and or audit procedure out of which a particular classification could be highlighted on the matrix.

6.5.5 Comparison of International Qualifications

Sometimes, there is a requirement for technicians and engineers from abroad to work in the UK or Ireland or vice versa. Although it is not practical to assess knowledge based on examinations passed or qualifications gained in other countries, a baseline can be established onto which a competence matrix can be populated. It is possible to set criteria for the minimum content for any knowledge-based learning. There are various organisations that can help in comparing the qualifications of individuals across Europe which are listed as follows.

UK NARIC – The National Agency responsible for information, advice, data and informed opinion on qualifications from outside the UK.

ECVET – The National Contact Point for England improving the mobility of those holding vocational qualifications across Europe.

ReferNet – The European Network offering comparable information on Vocational Education and Training across Europe.

Europass – The UK's National Europass Centre – removing barriers to work and study in Europe.

CPQ – The Centre for Professional Qualifications which provides advice and guidance on professional qualifications and their recognition in the UK, EU and globally.

All these organisations can be found on UK NARIC, website https://www.ecctis.com/.

6.5.6 Competence – Human Factors

Whenever we introduce people into risk management processes, we first have to consider that they sometimes will not behave in the way that we would want them to. We all make mistakes, but sometimes human failure can be deliberate which leads to non-compliance or it can be inadvertent which leads to errors. Whilst human factors are a huge topic on its own, I have devised some suggestions that may help when managing arc flash risk as follows.

Non-compliance is sometimes a cultural issue, or it can be an individual refusal to follow rules and procedures. In either case, it is a problem that must be faced head on and the reality has to be taken into account when devising risk control measures. Cultural issues are sometimes difficult to change as they inevitably come from the leadership within the organisation. It is at management level where we need to start in changing culture and to get agreement at the highest level through awareness and policy changes. It is then important that the workforce is involved in any changes to procedures which includes any risk control measures for arc flash. This is where awareness of the arcing hazard will be most effective. When and if it gets to protection by PPE, Chapter 7: Protection gives further advice on the involvement of the workforce in the provision of workwear, etc... We can also encourage reporting of non-compliance (violations) to make it unacceptable, akin to drink driving for instance. Where we know that there could be deliberate non-compliance, then targeted audits and raising the likelihood of sanctions must be considered.

In the case of inadvertent errors, we need to be aware of the need for ongoing training and competency assessment. Other factors that may lead to inadvertent errors are distractions and this is where control of the working environment is important, this is dealt with later.

Time pressure and tiredness particularly towards the end of a shift can be a vulnerable time. Another surprising common factor that I have come across on more than one occasion, is carrying out high risk activities out of hours that are on normal days off. Allow me to paint a picture of a non-shift working team who have been brought into work on a Saturday morning. Not wishing to stereotype, maybe Friday night is their usual night out and they will not be as bright on a Saturday morning. Maybe it is the football game on a Saturday afternoon so the time pressure is self-generated. Their performance would definitely be affected in contrast to if these factors were not in play.

Finally, I think that we also need to always be vigilant about the possibility of morning after intoxication through alcohol or drugs. Not a good mix when working near energised conductors.

6.5.7 Competence – Further Reading

Clearly, the competence of persons involved in live work will vary widely between industries and organisations. For instance, the requirements for someone involved in the manipulation of energised conductors by hot stick or live jointing is hugely different from a maintenance electrician undertaking diagnostic testing on a production line. It is not possible to cover all levels of work and supervision, but I can recommend the following comprehensive guide which was written by experts in live working across Europe. It is available to download free of charge from the International Social Security Association (ISSA) website. Guideline for Assessing the Competence of Persons involved in Live Working published by the International Social Security Association (ISBN-Nr. 3-9807576-6-8).

6.6 New Work and Alterations

There needs to be control over new work and alterations to electrical systems. It is often the case that the design of new installations or additions are left to the discretion of contractors even in medium to large facilities. Paperwork such as test sheets and minor works certificates are often not being handed over or even completed for modified electrical installations. This results in electrical installations that are a mixture of design specifications.

There needs to be a procedure that will ensure that any installation work is properly specified, designed, approved and installed to ensure consistency across the facility. The process will also deal with the inspection and testing, some of which may be on energised equipment and also the completion and handover.

6.7 Approval of Electrical Test Equipment & Tools

There needs to be control via a formal approvals process for all electrical test equipment, insulated tools, locks, labels, insulating tools, matting, shrouding and electrical PPE. This will require control over all such equipment on site and also equipment used by contractors. Incorrect test equipment is frequently used by the incorrect individuals and/or incorrect circumstances. For instance, there are often concerns as to the suitability of instruments to verify that circuits are dead. Engineers (including multi-skilled engineers who may or may not have electrical qualifications) use the voltage setting of digital multi-meters for testing for dead. The leads are often unfused, and this does not comply with the expectations of the enforcing authorities or guidance. Many times, multi-meters and leads that I inspect are not rated correctly for over voltage protection and had excessive visible tips on the probes. A very good reference on the subject is available as a free download from https://www.hse.gov.uk/ called Guidance Note GS38 (Fourth edition) on electrical test equipment for use on low voltage electrical systems. A suggested structure for the approval of tools and equipment is as follows.

Generic List. Create a generic approval list which will give standards to which all equipment shall be built. A generic list provides a specification for equipment where external contractors will be able to check their own equipment to ensure compliance with company standards.

Specific List. The next stage is to formulate a data base of specific list for equipment which is purchased by the company. Specific lists will describe the equipment, model number, manufacturer/supplier and other relevant information.

New Equipment. Next, create a process for the approval of new equipment so that items not listed on the specific list can be properly considered and control of equipment coming onto site is assured.

Specialised and Hired in Equipment. Finally, a process for specialised or hired in equipment, for example highly specialised for a particular task, such as harmonics monitoring equipment can be given a temporary approval by the site duty holder.

Maintenance and Calibration. Users must ensure that all specialist equipment is in good condition and correctly maintained and where necessary calibrated in accordance with the manufacturer's requirements.

PPE, Safety signs and Labels. Specialist electrical personal protective equipment such as shock and arc flash protection, approved safety signs and labels such as danger & caution notices and field marking labels should have a record kept in the specific list. The labels should include an image and approximate dimensions.

Audits. All equipment covered under the approval of tools and equipment should be audited on a regular basis, to ensure that it complies with the approved list of equipment. Any equipment that does not comply should be removed and destroyed.

6.8 Thermography and Partial Discharge Testing

Thermal imaging and partial discharge testing are classed as condition monitoring to detect impending failure in components and connections. This necessitates that the equipment is monitored in service and energised. Both techniques require some specialism particularly in the interpretation of results. Insurance companies sometimes provide discounts for companies that undertake thermal imaging mainly to reduce the risk of losses through electrical fires. Other insurance companies will insist on measures such as thermography. In addition, discovery of faulty joints or contacts could prevent overheating/arcing that could lead to a dangerous arc flash.

Can the procedures be carried out safely? The answer is yes, providing that a risk assessment which explores the alternatives has been properly implemented. As far as arc flash risk is concerned, this necessitates the predictive techniques always with a goal of prevention as first choice.

Thermal imaging requires that panel doors need to be opened in order to get accurate measurements. Sometimes internal shrouding may interfere with the thermographic images and I have come across situations where individuals may resort to the removal of shrouding. This obviously gives an extra order of risk to the operation and needs very careful consideration by the electrical duty holder. Among the alternatives to the opening of panel doors onto energised equipment there are:

a) Viewing windows which are mounted in the panel doors. These allow thermal imaging to occur without opening doors onto energised conductors.
b) Ductor testing which uses a high current digital micro-ohmmeter to measure very low resistances in circuit breakers contact resistance and joints in switchgear and busbar systems.
c) Joint inspection and tightening regimes as part of routine maintenance.

6.9 Risk Assessments – Practical Approach

Risk Assessments were spoken about in general terms in Chapter 3: Risk Assessment and the Four P Guide. What follows is some practical guidance on how risk assessments may be carried out on low voltage equipment which is where the majority of interactions with energised equipment occur.

Risk Assessments need to document the hazard severity and the risk control measures. Wherever possible, these should be dynamically produced, and task based. In other words, be available to the person carrying out the work who will reassess such things as environmental conditions and equipment state.

6.9.1 Can the normal policy of dead working be carried out?

Work on energised circuits must always be subject to a suitable and sufficient risk assessment. A key part of the risk assessment is the decision to proceed with live work rather than to have equipment dead and isolated for the duration of the work. To help with that decision, I have reproduced guidance from the excellent document, Electricity at Work – Safe Working Practices HSG 85 published by the UK Health and Safety Executive under Open Government Licence v3.0. I can thoroughly recommend the document which can now be obtained free of charge from https://www.hse.gov.uk. I have used the guidance as a framework not only to methodically breakdown the decision but establish the precautions that should be in place particularly in respect of arc flash but also for other electrical hazards. The following illustration, Figure 6.5, describes the process of decision making in greater detail.

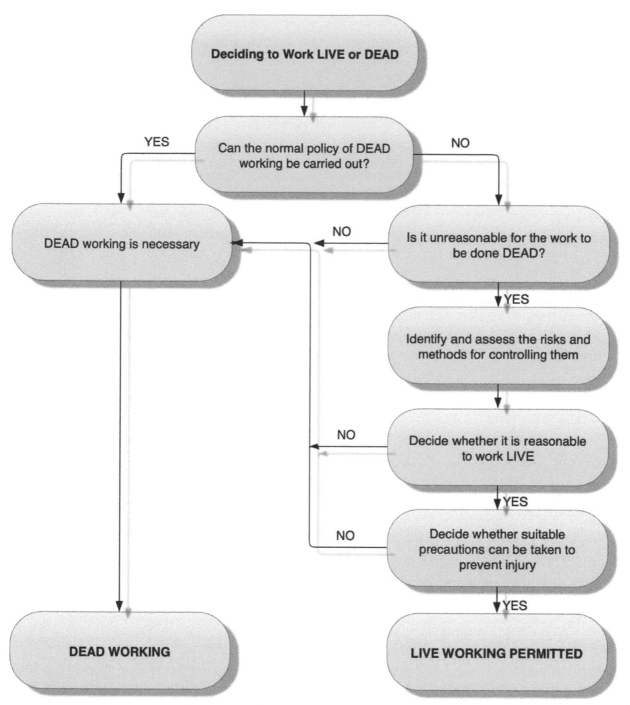

Deciding to Work LIVE or DEAD

Can the normal policy of DEAD working be carried out?

YES → DEAD working is necessary

NO → Is it unreasonable for the work to be done DEAD?

NO → DEAD working is necessary

YES → Identify and assess the risks and methods for controlling them

Decide whether it is reasonable to work LIVE

NO → DEAD working is necessary

YES → Decide whether suitable precautions can be taken to prevent injury

NO → DEAD working is necessary

YES → LIVE WORKING PERMITTED

DEAD WORKING

LIVE WORKING PERMITTED

Figure 6.5

Work on or near live exposed conductors should rarely be permitted. Many accidents to electricians, fitters, technicians and engineers occur when they are working on equipment that could have been isolated. You should plan and programme the work to allow all jobs to be carried out where possible with the equipment dead. Three conditions must be met for live working to be permitted where danger may arise. If just one of these conditions cannot be met, live working must not be permitted, and dead working is essential. The assessment procedure illustrates this. The conditions are:

a) it is unreasonable in all the circumstances for the conductor to be dead; and

b) it is reasonable in all the circumstances for the person to be at work on or near that conductor while it is live; and

c) suitable precautions (including, where necessary, the provision of personal protective equipment) have been taken to prevent injury.

6.9.2 Is it unreasonable for the work to be done DEAD?

The following examples are taken from HSG 85 to illustrate some circumstances where it is unreasonable to make equipment dead because of the difficulties it would cause:

- It may be difficult, if not impossible, to commission a complex control cabinet without having it energised at some time with parts live (but not exposed so that they may be easily touched);
- It may not be technically feasible to monitor the operation and performance of a control system or to trace a malfunction of such equipment with it dead, i.e., fault-finding;
- A distribution network operator (DNO) needs to connect a new low-voltage service to an existing main, but it might be unreasonable to disconnect many customers. In recognition of the dangers associated with live working, the DNO must have a very strict code of safety rules and procedures to prevent injury;
- Switching off a system, such as the supply to an electric railway track, to carry out maintenance or repair work may cause disproportionate disruption and cost.
- Identify, assess and evaluate the risks and methods for controlling them.

If you have decided that it is unreasonable for the work to be done dead, a risk assessment is necessary. The risk assessment must cover the work on or near the specific equipment and it must be carried out by someone with comprehensive knowledge and experience of the type of work and the means of controlling the risks. If live working can be justified through a rigorous test of reasonableness in conditions a) and b), judgements must be made about suitable precautions against electric shock and the effects of electrical flashover to satisfy the requirements of condition c). Work (which is not confined to just electrical work but includes any work activity) is permitted only if conditions a) and b) and c) are satisfied.

6.9.3 Identify and assess the risks & methods for controlling them

The first stage of the risk assessment is to identify the hazards, and although not exhaustive, the following highlights some of the more common ones.

- Electrical shock from passage of current through the body, leading to death or injury due to cardiac arrest, shutting down of the respiratory system and heart fibrillation at extremely small currents. This is typically greater than 30mA on alternating current systems. Muscle contraction can occur.

- Internal burns where the current involved is in the order of amperes. The destruction of body tissue and internal organs can occur with the passage of current at around one ampere.
- Arc flash injury can include external burns (i.e., severe burns to the skin), internal burns and intoxication from inhaling hot gases and vaporised metal, hearing damage, eye damage and blindness from the ultraviolet light of the flash as well as many other devastating injuries. Depending on the severity of the arc flash, an explosive force known as an arc blast may also occur and could cause the propulsion of molten metal, equipment parts and other debris.
- Ignition of flammable materials from electrical arcing which could be the clothing that an electrical worker is wearing. Serious third-degree burns have occurred to workers wearing flammable clothing which could become a hazard irrespective of PPE considerations.

6.10 Dynamic Risk Assessment Checklist

Having identified the hazards arising from the live work, you then must decide how likely it is that harm will occur and the severity of injury that might occur. The following is a check list, Figure 6.6, of issues to consider that may impact on the likelihood of the hazards causing harm.

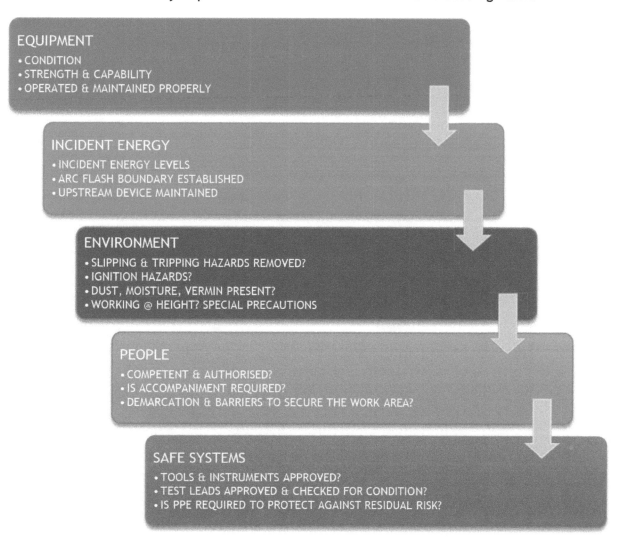

Figure 6.6 Risk Assessment Checklist

6.10.1 Equipment

ALWAYS ASSUME THAT EQUIPMENT OR CONDUCTORS ARE LIVE UNLESS PROVEN DEAD!

Has the equipment been checked and is it in a safe condition? Check whether the equipment to be worked upon has been examined and in a safe condition for work. Live work should never be permitted where there are any doubts about the safety of cables and electrical equipment being worked upon or adjacent to those being worked upon. The examination can be visual but also using other senses. Smell and hearing can detect burning or electrical discharge. Signs of vermin, birds or organic growths inside switchgear or water ingress is a definite prompt to stop and investigate when the switchgear can be made dead and isolated. Approaches should never be made to damaged cables. Check that the equipment doors, panels and covers are closed and secured and there is there is no evidence of impending failure.

For new equipment, the person undertaking the risk assessment should be confident that the equipment has been professionally installed in accordance with the design specification and manufacturer's instructions. For existing equipment, in addition, a judgement should be made that it is being maintained and is operating in accordance with the design parameters. Is the equipment over stressed? An arc flash study if properly conducted will provide a single line diagram, a protection coordination study and assure that the strength and capability of the equipment is adequate.

If the equipment is in a safe condition the next step is to consider whether the equipment is finger safe. If the equipment is not finger safe, can measures to prevent contact with live parts be implemented? The term "Finger Safe" is defined as no exposed live parts that can be accessed by solid objects greater than 12.5mm as given by IP rating IP2X.

Do not rely purely on the original specification of the equipment. Insulation is often removed and not replaced. If it is not finger safe, or other measures cannot be introduced to prevent contact with live parts then proceed no further.

If a maintenance program has been implemented as suggested earlier in this chapter, there may be established risk control measures to be adopted when working on a particular piece of equipment. These controls may be displayed in the form of labelling on the equipment which should be taken into account in this risk assessment process.

6.10.2 Arc Flash

A major objective of this guide is to provide the predictive tools to determine the arc flash hazard and the means to prevent it causing harm. As part of the risk assessment, the degree of harm may be predicted and quantified in terms of incident energy and distance. This is very useful, particularly on low voltage equipment where interactions with energised equipment and cables are countless and the relative severity of the hazard has little to do with nominal current ratings or available prospective short circuit current. (See Chapter 12: Myths and Mistakes)

By using this guide and the tools for determining the incident energy, a quantitative analysis can be made about the severity of a possible arc flash which will have a bearing on this decision-making process and also provide the specification data on any protection requirements. The conclusion may be that the incident energy levels are too high and cannot be suitably controlled but experience shows that preventative measures can be formulated in the majority of cases at low voltage.

Using the IEEE 1584 Guide for Performing Arc Flash Calculations, the incident energy level at a distance to the worker is determined. The distance is normally to the head and torso but the calculators with this guide can easily be used to determine the incident energy to other parts of the body and particularly the hands. Where the arc flash hazard is dependent upon tripping time for the upstream protective device, there needs to be confidence that this has been adequately maintained?

Boundaries

It should be ensured that adequate clearances are established and maintained when working on or near to live equipment. Annex A of BS EN 50110-1:2013, Operation of Electrical Systems, gives guidance for shock protection distances in air for working procedures. As we are speaking about low voltage systems here, any contact either by the worker or tools/instruments is classed as live working. Any encroachment below 300mm from the live part below 1000 volts, is classed as "work in the vicinity".

Using the IEEE 1584 Guide for Performing Arc Flash Calculations, the arc flash boundary can be determined. This is defined as a distance from a prospective arc source at which the incident energy is calculated to be $5.0J/cm^2$ ($1.2\ cal/cm^2$) The significance of this figure, is that this amount of thermal energy is calculated to have a 50% chance of a second degree burn to bare skin if sustained for 1 second. This value is used by many standards as the benchmark that defines protection against the thermal effects of arc flash and the threshold of a zone. For instance, working on or near energised equipment within the arc flash boundary will require the wearing of arc flash PPE. The calculation of the arc flash boundary is very useful regardless of whether prescriptive standards may be applicable in determining the level of harm from the arc flash hazard at a determined distance. For further information, See Chapter 14: Hazard and Severity Calculators.

6.10.3 Environment

Is access and space adequate? Establish whether the access and space in front of the equipment is adequate to allow the worker to pull back from the conductors without hazard. Refer also to the earlier arc flash section on the shock and arc flash boundaries.

The work area should be clearly defined, with no tripping & slipping hazards and with good means of escape and illumination. Simple barriers and signs can be erected for the demarcation of work areas to keep unauthorised staff away and also to protect electrical workers from interruptions at times when they need to concentrate.

Is lighting adequate? Also, it is important to check whether lighting levels are adequate for work. Use of additional lighting is essential where required. Further details on lighting at work can be found in HSG38 Lighting at work which is free to download on https://www.hse.gov.uk/.

Are hazardous conditions present? Check to ensure that the immediate environment is free from water or dust. A hostile or wet environment will significantly increase the risk and severity of electric shock and should therefore be subject of special consideration to control the risks.

Ensuring there is no possibility of an ignition hazard due to sparks is crucial. If there are signs that there is a possibility of an ignition hazard, take precautions to remove the hazard before proceeding with the working in accordance with company procedures.

6.10.4 People and Safe Systems of Work

As described previously, it is essential to check whether the workers are competent for the task. In the context of live work, technical knowledge or experience means that the person should be properly trained and assessed in the techniques being employed but the person must also understand the hazards from the system and be able to recognise whether it is safe for the work to continue. The person should, as a minimum, have a specific authorisation to carry out live working tasks under any electrical safety rules.

Is the work to be carried out at height? Working at height whilst carrying out live work is always a special case for consideration for two reasons:

a) Electric shock or arc flash to a worker at height can bring about a fall with obvious consequences.
b) An arc flash incident whilst working at height may mean that the worker cannot move out of the way because of the limited working space on access equipment. This may be the work platform of a scaffold or a mobile elevated work platform.

If the work is to be carried out at height can risk control measures to prevent shock, burns and falls be put in place?

Consider the use of protective screens or insulation to prevent simultaneous contact with live parts at different potentials. These are commercially available and an example of temporary insulation together with insulating fasteners is given in Figure 6.7.

Establish whether accompaniment is required. Anyone undertaking work on or near energised electrical conductors will nearly always require

Figure 6.7 Image courtesy of Boddington Electrical

some form of accompaniment by someone who can give assistance in an emergency. The emergency backup may be for the accompanying person to be able to make the system safe and to provide a medical response as may be necessary. This implies a degree of competence such that the accompanying person can assist without danger to themselves or others. A requirement for a second person is to implement and maintain safe working procedures such as by preventing distractions and encroachment from non-authorised personnel into the working area.

Provide and use the correct personal protective equipment (PPE) to reduce the risk of contact with live parts or earth through insulating gloves and matting. If there is a risk of burns from arc flash that cannot be eliminated by other means, use adequately rated, thermally insulating, flame-resistant PPE. Personal protective equipment (PPE) is dealt with separately in Chapter 7: Protection.

If measures to prevent contact with live parts can be implemented, are tools, instruments and leads checked fit for purpose? Tools and instruments must be of the correct duty rating and their condition must be checked especially test leads. It is important that correct instruments and leads should be selected and in particular the correct over voltage installation category. The wrong meter and leads can increase the chances of electric shock or the instigation of an arc flash due to transient over voltage. Only approved test equipment and leads should be used, and the leads should be fused and have insulated probes.

I have come across many organisations who adopt a permit to work for live work. Indeed, the Standard for Electrical Safety in the Workplace NFPA 70E in the USA mandates the use of an Energized Electrical Work Permit. BEWARE! It should be noted that the use of an energized electrical work permit would not be acceptable in the UK as electrical permits are restricted to dead working only. To quote HSG 85 Electricity at Work – Safe Working Practices "An electrical permit-to-work is primarily a statement that a circuit or item of equipment is safe to work on, it has been isolated and, where appropriate, earthed. You must never issue an electrical permit-to-work for work on equipment that is still live or to authorise live work."

However, all the information for non-routine work should be documented as special circumstances in the risk assessment and if necessary, a sanction for testing/live work can act as a supplement to give greater authority.

6.10.5 Decide whether it is reasonable to work live!

"The risk assessment should inform managers and supervisors whether it is reasonable in all the circumstances to work live. The decision should not be taken lightly. At this stage the economic and operational factors should be evaluated against the risks involved before making a decision, bearing in mind that the risks associated with working live can be very serious. Minor inconveniences arising from working with the equipment dead, sometimes arising from commercial and time pressures, will very rarely outweigh the risks associated with live work". (Source: HSG 85 Electricity at Work – Safe Working Practices)

6.10.6 Decide whether suitable precautions can be taken to prevent injury!

Once the above requirements have been met, the third condition for the justification of live working must be met which is: suitable precautions (including, where necessary, the provision of personal protective equipment) have been taken to prevent injury. At this stage, the hazards and the precautions to prevent injury should all have been identified and recorded in the risk assessment as detailed in previous paragraphs.

6.10.7 FINALLY - Proceeding with Work

After all the above steps are satisfied, implement safe working and ensure adequate monitoring and supervision. Things change in workplaces, and risk assessments need to be reviewed on an ongoing basis. Consider a "toolbox talk" to run through the requirements of the risk assessment and ensure that these are understood by those undertaking the work.

In addition, the worker must:

- Make sure that any special equipment and PPE is properly used and maintained.
- Pay attention to housekeeping and ensure that the work area is organised and tidy.
- Keep the duration of any live work to a minimum.
- Store tools correctly – horizontal surfaces and projections inside control cabinets should not be used – and ensure that objects such as tools and bolts cannot fall onto exposed live parts.
- Maintain tools and test equipment in good condition and replace them if damaged.
- Avoid lone live working at any stage and make sure that any accompanying person is competent.

6.11 Testing for DEAD

A core principle of most electrical safety programmes is dead working and to always try to avoid live working. Why I emphasise the avoidance of live working, is that sometimes, there can be danger in making equipment dead, or putting equipment in an electrically safe condition.

I recall, many years ago, a utility company repairs electrician being called out to look at a low voltage cable which had been severed by excavation in the street. The current carrying conductors in the severed armoured cable were clearly visible and the electrician jumped into the trench to see if they were still live. The conductors were still live, but the act of prodding them with a test lamp initiated an arc flash incident in which he received some burns. A footnote to this incident is that the same electrician did a very similar thing some years later, but this time was very severely burnt from the resulting fireball and was hospitalised.

Remember!

TREAT ALL ELECTRICAL EQUIPMENT AND CONDUCTORS AS LIVE UNLESS PROVEN DEAD

So, with that in mind, how shall we safely test for dead. Whilst the following is not meant to be a definitive guide to isolation practices or lock out tag out procedures, it is a practical approach to ensure that the process of testing at low voltage will not create danger from arc flash. This is to be taken as a whole and no one measure should be taken in isolation.

STEP 1. COMPETENCE - Designate the person who will have responsibility for the isolation. We will call him/her the designated person here, but that term can be changed as long as it is clear, written into the safety rules and does not clash or be confused with any other terms in use. Make sure that the designated person possesses the competence required to undertake the isolation. By that I mean a person who has the skill, knowledge, attitude, training, possess the experience to undertake the work and has the ability to recognise the extent of their own competence and leading them to take appropriate action. Specific requirements over and above general electrotechnical knowledge are that he/she will have an adequate understanding of the system to be worked on and practical experience of that class of system, an understanding of the hazards which may arise and the precautions which need to be taken and an ability to recognise at all times whether it is safe for work to continue. Sadly, the repairs electrician that I spoke about earlier failed on all three counts.

STEP 2. IDENTIFY. The designated person must correctly identify and make dead and Isolated from ALL points of supply. This will include alternative feeders and emergency standby arrangements such as generators and uninterruptible power supply systems. Correctly identifying circuits will take away the trial-and-error approach and reduce the risk.

Do the detective work. Reference to current drawings (taking care to allow for inaccuracies) and records alongside circuit and equipment labelling is a good start. Consult the electrical system owner or experienced maintenance crews whenever possible. Use physical inspection such as tracing out circuits to determine the correct isolation points relating to the work required. If the equipment is a distribution board, identify an outgoing load that is non-essential such as a lighting circuit. When switching and verifying it can be another step in positively identifying the points of isolation by observing simultaneous extinguishing of lighting circuits through the switching operation. A similar approach can be made with panel control lamps if the piece of equipment is a power switchgear and control gear assembly. This may sound like stating the blindingly obvious, but believe me, having audited individuals whilst they are undertaking isolation procedures, it surprises me how often these simple steps are omitted.

Testing because of inadequate records. Where testing is carried out to locate circuits due to inadequate records or circuit labelling then this should be considered as live working.

Unauthorised modifications or circuits. Finally, always be open to the possibility that there may be unauthorised feeders and/or modification that are not detailed in circuit diagrams and records.

STEP 3. DISCONNECT.

The usual means of disconnection of circuits at low voltage are:

a) the withdrawal of fuses and/or links
b) operation of miniature and moulded circuit breakers
c) opening and isolation of air circuit breakers
d) opening of switch fuses
e) opening of isolators
f) physical disconnection of conductors
g) withdrawal of a plug from the socket outlet
h) third party isolation via a utility company

Notes.

Reduce connected load. Before operating isolators or removing fuses reduce as much of the connected load as possible.

Physical disconnection of conductors. This can only to be carried out dead, but this may be the only option in certain circumstances. For instance, if the disconnection device is not rated as an isolator or that the point of isolation cannot be secured with a padlock or other device.

Visual inspection of isolator blades. Make sure, wherever possible and visible, to check and to ensure that all the blades have opened in switch fuses and isolators. I have seen it first-hand where the operating bar has broken leaving a blade in place and energised. I also remember a recall notice for a popular brand of devices for that same fault.

Circuit Breakers in the tripped position. Miniature and moulded circuit breakers can never be used for isolation in the "tripped" only position. This is one of the mistakes that led to the incident mentioned in my introduction to the guide.

Storage Devices. Ensure that storage devices such as capacitors are fully discharged. Capacitors can store huge amounts of energy that can cause a serious blast hazard. If you have concerns about the capacitors, I can recommend the 2021 version of NFPA 70E which has the subject covered in some detail in Informative Index R. It is free to view, and I have given details of the document including caveats in Chapter 16: Rules, Codes and Legislation.

STEP 4. VERIFY INCLUDING TESTING.

The whole point of steps one, two and three is that verification will be on dead conductors. We have already reduced the risk from arc flash if these steps have been carried out diligently. But should there still be a residual risk the following will serve to reduce it to as close to zero as possible. Know what the incident energy level is and take a decision on how to proceed. Decisions should be based upon risk assessment especially in respect of whether to deploy PPE or not. There is a good deal of guidance on this elsewhere in the guide especially in Chapter 4: Prediction and Chapter 7: Protection but what I would suggest is a minimum of hand protection regardless of the incident energy levels for low voltage testing. So, in other words, insulating gloves and leather over gloves. In Chapter 12: Myths and Mistakes, I give an example of an electrician who was hospitalised after suffering burns to his hand. I put objections to wearing gloves for testing in the same category as not wearing a mask. There may be some inconvenience, but loss of dexterity is in my view, overblown.

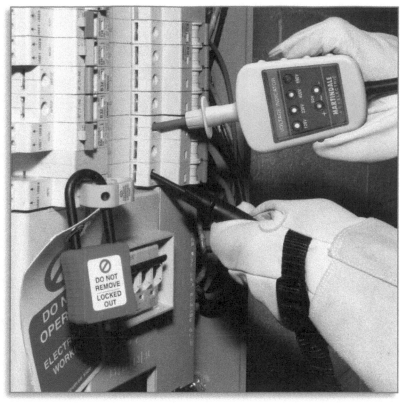

Figure 6.8 Image courtesy of Martindale Electric

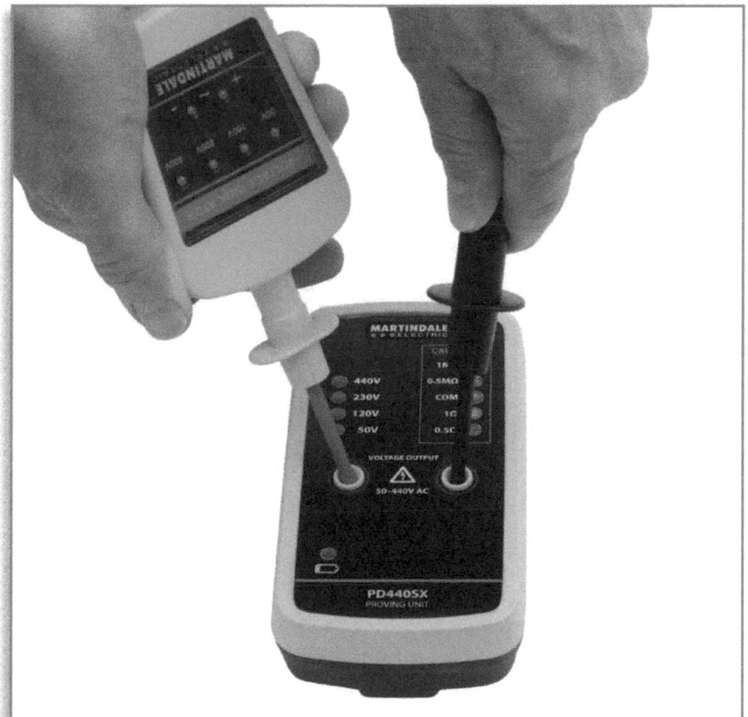

Figure 6.9 Image courtesy of Martindale Electric

Use the correct equipment. I would suggest that multi-meters should not be used for testing for dead. Correctly calibrated proprietary voltage indicators, with fused leads and matching test units (Figure 6.9) should be used to check for the absence of voltage.

Non-contact voltage indicators (Figure 6.10) are a good idea providing that they are **NOT** relied upon for the final proof of dead conductors. Look for devices that have self-proving circuits and can give an indication if they are faulty. They may obviate the need for PPE for instance as used as part of the risk assessment given earlier. In addition, they can be used to look for rogue circuits. For example, in a control panel or composite switchboard, cables may have been drawn in that are supplied from another source. The panel may have been used as a trunking so there may be insulated but live cables in the working area. Make sure that you fully understand what the non-contact voltage indicators are capable of from the manufacturer's data and instructions especially voltage ranges.

Figure 6.10 Image courtesy of Martindale Electric

Test ALL Poles of Supply. The arc flash incident mentioned in the introduction was a classic case of not testing all poles of supply between current carrying conductors and then each current carrying conductor and earth including neutrals. There was a failure to test to earth which would have detected that the red phase contact on the controlling moulded case circuit breaker was welded closed whilst the other two phases were open.

Test the lowest rated circuits first. For instance, if the piece of equipment is a power switchgear and control gear assembly then test at the control fuses or an outgoing protective device.

Lower upstream device settings. If the incident energy levels are still very high, consider lowering the upstream protection settings. This does need care to make sure that appropriate settings are reapplied upon completion.

STEP 5. SECURE.

ALL practical steps to prevent re-energisation should be taken with the goal that there is no risk that circuit conductors and electrical equipment can be made live, whilst work is in progress. Some measures can include:

a) The retention of fuses, links and withdrawable circuit breakers and application of lockable blanks and retention of keys.
b) Disconnection of circuit conductors.
c) Locking off points of isolation such as circuit breakers, isolators and the retention of keys.
d) Removal and retention of castell keys where these mechanically interlock with switches or circuit breakers ensuring that spare keys are not available.
e) Specially designed locked facilities, e.g., lockout boxes.
f) Use of properly designed earthing devices. (See below)

Earthing. Applying circuit main earths where they are available is mandatory. For other situations this is an option for the designated person to consider bearing in mind that control needs to be exercised in their application and removal.

Lock and Label Points of Isolation. Secure all points isolation using a lock and unique key. Post caution notices at all points of isolation and also on any earths that have been applied. Make sure that all notices are removed on completion.

Labelling adjacent equipment. If there is live equipment in the vicinity that could be mistaken for the equipment to be worked upon, then apply a label. A "danger live" notice will usually suffice, the presence of which should be noted on any permits that are in use and they must be removed on completion.

Third Party Isolations. An isolation from a third party such as a utility company, needs to be thought through carefully as the designated person for securing the isolation should always have control. There are various ways in which this is achieved, and this is how I have gone about it in the past. Firstly, try to accompany the person from the third party who is undertaking the isolation and be assured from discussion and observation that this is the correct point of isolation. Secondly, ask that person to accept a means of securing that point of isolation such as a lockout box with your lock on it or a multi hasp device. Finally, ask for a written statement of the isolation. This will not be a permit to work as the third party is not responsible for your working party (there could be other sources of supply outside his/her control), and it is likely to be an isolation and earthing certificate. Most utility engineers that I have worked with will cooperate in assisting and assuring you in order to help you to fulfil your responsibility to set up a safe system of work.

Generators. Work on generating and motor driven plant is not to be carried out unless the machine concerned has been securely isolated from all possible rotational driving forces. Particular attention is to be given to the isolation of the auxiliary circuits, controlling the automatic start sequence of any prime mover associated with a generator.

Distance. Finally, putting any distance between the test site and the person is beneficial because of the exponential reduction of incident energy as discussed previously.

STEP 6. AUTHORISE.

Even though we are speaking here about low voltage isolations, an electrical permit to work system is beneficial for any complex commercial or industrial site. The following is meant as a brief overview only, but a permit system should be carefully considered and be drafted into electrical safety rules. A permit to work should always be issued where the designated person undertaking the isolation is not the one who is going to do the work and self-permitting is to be discouraged. The specific circumstances where a low voltage permit to work can be issued are:

a) Where the designated person is isolating for others.
b) For multiple working parties.
c) For complex isolations where there may be more than one point of supply.
d) Where third parties such as a utility company are involved in isolating their equipment to provide you with a point of isolation.
e) Switchgear with automatic changeover or remote-control switchgear and equipment controlled by phase failure relays.
f) Extensions to switchboards or control panels.
g) Isolations from Generators and UPS equipment.
h) Stored energy systems (either mechanical or electrical).
i) Isolations lasting more than one shift.

The list is not exhaustive and there may be other situations requiring the use of permits.

6.12 Live Testing, Fault Finding and Diagnostics

Testing for fault finding, diagnostic or commissioning purposes on live equipment, is live work and precautions should be taken to reduce the risk of arc flash (and electric shock of course) to as low as is reasonably practicable. If workers are routinely testing on very high incident energy circuits, I would suggest that Chapter 5: Prevention should be revisited as a matter of urgency. The use of extra low voltage and/or segregation of control circuits from power circuits is always the desired option and if that is not possible because of legacy equipment design then the following checklist may help to prevent accidents.

6.12.1 Live Testing Risk Assessment Checklist for Low Voltage Equipment

Planning for Safety. If the live testing is for diagnostics or fault finding, I would suggest that an equipment audit will be a good start. There is an example of a control panel audit form which can be used as one part of the audit step in Chapter 9: Electrical Duty Holders. It will be an opportunity to not only deal with the arc flash risk but also very effectively take stock of equipment which is likely to have the greatest number of interactions and therefore risk in terms of likelihood of electrical incidents.

Incident Energy Level. As described earlier, knowledge of what the incident energy level is will allow for a decision on how to proceed. Decisions should be based upon risk assessment especially in respect of whether to deploy PPE or not. As I stated under testing for dead, I would suggest a minimum of hand protection regardless of the incident energy levels for low voltage testing comprising of insulating gloves and leather over gloves. Consider also the routine wearing of face and eye protection for live testing.

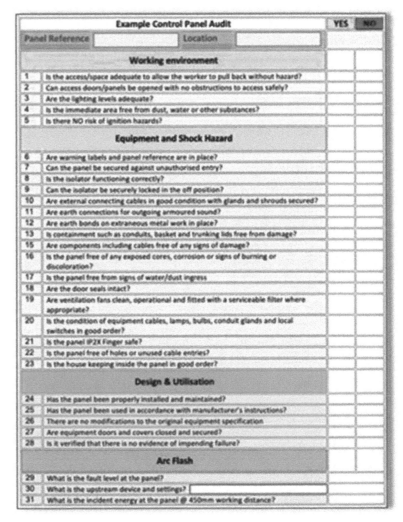

Figure 6.11 Example of a Control Panel Audit Form

Competence. A high degree of competence is required for live testing and although I have said it previously under dead testing, it is worth repeating. Competence means that the worker must have the skill, knowledge, attitude, training, possess the experience to undertake the work and has the ability to recognise the extent of their own competence and leading them to take appropriate action. Specific requirements over and above general electrotechnical knowledge are that he/she will have an adequate understanding of the system to be worked on and practical experience of that class of system, an understanding of the hazards which may arise and the precautions which need to be taken and an ability to recognise at all times whether it is safe for work to continue. What I would add here is that multi-skilled "technicians" are often used for fault finding on electrical equipment whose first calling may not be electrical. I can relate many experiences of this approach which are quite uncomfortable, and this is a route that needs skill and care that such individuals have the necessary electrotechnical knowledge to avoid endangering themselves or others.

Secure the work area. Sufficient space should be provided and maintained around the equipment to permit safe testing. Make sure that tripping and slipping hazards are removed and the work area is under the complete control of the person undertaking the tests. This could be by accompaniment and/or the area is demarcated or defined with barriers.

Work at height. Both the arc flash and electrical shock hazards could be made much more deadly when testing or carrying out any other work on or near live equipment at high level. Special consideration should be given to working at height in the risk assessment with a view to avoidance. Even minor arc flash incidents at height can cause falls or in one case that I know about, injury by jumping from a mobile elevated work platform.

Keep the work to a minimum. The work should be kept to a minimum and the panel doors and access should be secured at breaks or at the end of a shift.

Illumination of work area. Lighting levels should be adequate for the person carrying out the work. Natural light is preferable but permanent artificial lighting improvements should be implemented. Handheld torches are not recommended and should be avoided.

Correct test equipment. The standard IEC 61010-1 Safety requirements for electrical equipment for measurement, control, and laboratory use gives categories of hazardous transient impulse overvoltage protection in four categories. The closer to the source the higher the degree of protection is required. I would suggest that testing on anything that has the potential to cause serious injury from arc flash ought to be nothing less than Cat IV impulse voltage protection for industrial and commercial applications. I would also recommend good quality fused leads which are pictured in Figure 6.12.

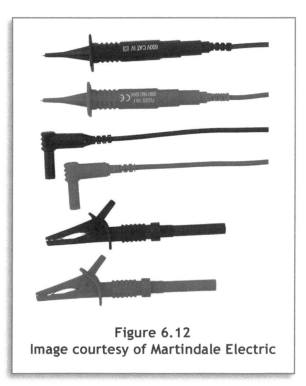

Figure 6.12
Image courtesy of Martindale Electric

Accompaniment. In high-risk situations accompaniment ought to be a consideration. An example may be where there may be difficulty in keeping people away from the work area. The person providing the accompaniment must be trained to be able to recognise danger and give assistance in the event of an emergency.

Prohibit removal of components. Removal of live components and the disconnection of live components should always be prohibited and made clear in the electrical safety rules.

Insulated Tools. At the age of 16 years as an apprentice electrician, I was given my first tool bag. It was full of brand-new high quality, fully insulated tools. It was not a signal to me that I would be carrying out live work but I did recognise the potential risk of using uninsulated tools anywhere near energised equipment. Even routine testing may include the use of tools for access and running adjustments. The standard IEC 60900 is applicable for hand tools for use up to 1000 volts AC and 1500 volts DC.

Remove jewellery. Metal jewellery such as watches, rings, bracelets, piercings, and neck chains should be removed before commencement of work.

6.12.2 Battery Testing

Work on large battery systems, even at extra low voltage can be hazardous and common injuries are arc flash burns from electrical melting of metallic objects inadvertently shorted across the battery terminals and also electrically ignited hydrogen/oxygen explosions. Consider the following non exhaustive options for work on or near batteries.

Incident Energy Level. Awareness of the arc flash hazard level is important and there is a hazard calculator provided for the assessment of the incident energy level at direct current. As always this should be taken into account in the risk assessment process in terms of overall risk and is not a direct route into PPE. This all depends on the task and the state of the equipment.

Competence. Workers who are engaged in the connection, disconnection and testing of secondary battery systems should be competent in all the ways previously expressed. In addition to the competence requirements outlined in live testing above, they will be persons who are considered to have the necessary experience, personal qualities and training to work safely and recognise the risk involved when working on secondary batteries. The training will include, where necessary, the precautions for non-electrical hazards such as the handling of electrolyte.

Personal Protective Equipment. Face protection and insulating gloves are worn regardless of incident energy. This is where care needs to be taken in the selection of PPE that cover all hazards and do not create a hazard. This is where manufacturer's handling instructions need to be taken into account especially where contact with liquid electrolyte may be encountered.

The work area is well ventilated. There may be a requirement for specialist aprons for the chemical hazards as an example, and as pointed out in the next Chapter 7: Protection, all personal protective equipment must be appropriate for the risks involved, without itself leading to any increased risk.

Other precautions. As described in live testing above; the extent of live working should be kept to a minimum, metal jewellery such as watches, rings, bracelets and neck chains should be removed before commencement of work and correctly selected insulated tools and instruments should be used.

There is an excellent free resource available on the Great Britain Health and Safety Executive site called "Using Electric Storage Batteries Safely". It can be downloaded via the link; https://www.hse.gov.uk/pubns/indg139.pdf

---

Learning Points

➢ The competence of the individual is of paramount importance whether they are making dead or working anywhere near energised conductors.

➢ Having safe procedures, safe people and safe places will provide the building blocks to arc flash risk management and electrical safety.

➢ The legal requirement for safety signs is different in Europe. Safety signs are where there is a significant risk to health and safety that has not been removed or controlled by other methods. Also, the colours are totally different to the US.

➢ Live working permits are definitely not a good idea in the UK.

➢ Treat all electrical equipment and conductors as live unless proven dead.

Chapter 7

Protection

Where the risk cannot be controlled by prevention or where there is a residual risk of injury then it may be necessary to consider mitigation to prevent injury to the worker. The requirement for and suitability of mitigation techniques must form an essential element of any risk assessment. In Chapter 3: Risk Assessments and the Four P Guide, there was a hierarchy of risk controls for electrical flashover and the final risk control measure in the hierarchy is Personal Protective Equipment. It is fairly obvious that PPE is not mentioned in the principles of prevention given in Chapter 3: Risk Assessments and the Four P Guide as PPE alone will not prevent the accident. It is therefore seen as a last line of defence but where used properly, there is evidence that it has prevented injury to individuals. In this chapter, the main focus is on the selection of PPE for protection against arc flash and it will look at how this aligns with the risk assessment process. Although the intention is to inform about how arc flash PPE is selected and used, one objective is to simplify the selection process. In most situations, flame resistant PPE is surprisingly comfortable nowadays and can be worn as everyday workwear. If we follow the steps laid out in this guide so far, the residual risk should now be much less severe and hopefully minimised to a great extent. That means that should PPE be required as a final risk control measure, then this will be much lighter and less cumbersome. The wearing of PPE should be much more acceptable, not only in terms of user agreement but also in expense. Heavy duty PPE is expensive!

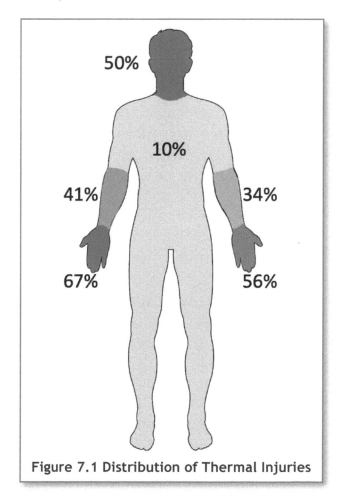

Figure 7.1 Distribution of Thermal Injuries

Non-flame-resistant clothing may ignite or melt at low incident energy values and once ignited will continue to burn after the electrical arc has been extinguished. Three seconds of burning material next to the flesh can result in serious full thickness burns. This means that ordinary clothing could actually become a hazard and for this reason it is well worth considering a policy of comfortable Flame Resistant (FR) clothing. There is research that shows that the hands, arms and face are the most commonly affected parts of the body in an arc flash incident. See Figure 7.1 (Source: Guideline for selection of personnel protective clothing when exposed to the thermal effects of an electric arc, International Social Security Association- Section Electricity, Koln, 2002) This is not surprising as these parts of the body are often exposed and are also much closer to the arc, special emphasis should be given to face and hand protection.

7.1 Standards for PPE in Europe

European Council Directive 89/656/EEC Use of Personal Protective Equipment covers the minimum health and safety requirements for the use by workers of personal protective equipment at the workplace. Laws in EU countries on the use of personal protective equipment (PPE) in the workplace are all based upon this directive. You can find the specific legislation in Chapter 16: Rules, Codes and Legislation for the UK and Ireland. Priority must be given to collective safety measures. PPE can only be used where the existing risks cannot be sufficiently limited by technical means or collective protection or work organization procedures. The employer must also provide the appropriate equipment free of charge and ensure that it is in good working order and hygienic condition.

PPE can only be prescribed after the employer has analysed and assessed the risks which cannot be avoided by other means. For arc flash this means an employer must consider other means of achieving safety prior to considering the use of PPE, such as the elimination of hazard, engineering controls and safe systems of work.

Issues of use, which must be part of the risk assessment process, must include ergonomics, sensory deprivation of user, continuing integrity of PPE and other injury mechanisms, including loss of hearing and sight. Heat stress is a particular concern with arc flash PPE as it may lead to fatigue at best and could lead to poor decision making, not a good thing when working near live conductors. This is one more reason to ensure that the risk assessment process is followed rigorously with the avoidance of over specified PPE. In addition, the health of the worker must be considered, bearing in mind that arc flash PPE will usually result in some restriction of the face. There could therefore be some physical and even psychological issues to take into account.

In addition to the above is the Regulation (EU) 2016/425 of the European Parliament and of the Council of 9 March 2016 on Personal Protective Equipment, repealing Council Directive 89/686/EEC. The regulation lays down requirements for the design and manufacture of personal protective equipment (PPE), which is to be made available on the market, in order to ensure protection of the health and safety of users and establish rules on the free movement of PPE in the European Union.

7.2 Selection of PPE

Article 5 of Directive 89/656/EEC clarifies that assessment of personal protective equipment shall involve an analysis and assessment of risks which cannot be avoided by other means. In respect of arc flash, to start off with a quantitative assessment based on incident energy will go some way to satisfying this requirement. Before choosing personal protective equipment, Article 5 asks us to assess whether the PPE that we intend to use will satisfy the following requirements.

(1) Personal protective equipment must comply with the relevant community provisions on design and manufacture with respect to safety and health. All personal protective equipment must:

- Be appropriate for the risks involved, without itself leading to any increased risk.
- Correspond to existing conditions at the workplace.
- Take account of ergonomic requirements and the worker's state of health.
- Fit the wearer correctly after any necessary adjustment.

(2) Where the presence of more than one risk makes it necessary for a worker to wear simultaneously more than one item of personal protective equipment, such equipment must be compatible and continue to be effective against the risk or risks in question.

The assessment to meet the above requirements involves a three-step process as follows: *(Note, that the italics are guidance on how that step can be implemented)*

> **(A) an ANALYSIS AND ASSESSMENT of risks which cannot be avoided by other means.**
>
> *HOW? By predicting the severity of the arc. See Chapter 14: Hazard and Severity Calculations. If the risk assessment has been carried out in accordance with four P's principle, we are left with a residual risk which will be as small as possible making the next steps easier and lead to a more comfortable and less expensive solution. At this stage we will have a numerical value for the incident energy at a working distance and also an arc flash boundary which will give a distance at which no PPE is necessary. It has to be remembered that different parts of the body will be affected in different ways. If you use the tools available with this guide, you will be able to instantly determine the incident energy at any part of the body in a dynamic way. For instance, the hands may be closer to an arc and have a much greater magnitude of incident energy. (Alternatively, if undertaking an assessment under the IEC64182-1-2 Box test method we will arrive at either an arc protection class (APC) 1 or 2)*

> **(B) the DEFINITION OF THE CHARACTERISTICS which PPE must have in order to be effective against the risks referred to in (A), taking into account any risks which this equipment itself may create.**
>
> *HOW? Whilst the arc flash attributes can be established by straightforward calculation there are other characteristics that need to be considered before we can go to the marketplace with a shopping list for suitable PPE. For instance, the PPE may be required in a clean room environment, or where there may be a biological hazard or outdoors or perhaps a very dirty*

area which may degrade or compromise the type of FR materials available. This needs to be identified as environmental considerations. Finally, the characteristics need to include all the other risks that have been identified previously such as ergonomics, compatibility with other protection and fit.

(C) **COMPARISON OF THE CHARACTERISTICS of the PPE available with the characteristics referred to in (B).**

HOW? CE marking is essential so should be the starting point when making comparisons in the marketplace. There have been great developments in the design of PPE systems in recent years and systems are available for parts of the body of environmental conditions. However, depending upon the environmental considerations and other risks, there may have to be compromises when considering the available choices. One of the issues that will crop up when choosing FR fabrics will be how the flame resistant qualities are achieved. There are permanently treated fabrics, predominantly cotton which is chemically treated and then those that are inherently flame resistant. There will be fabric weight, comfort and life expectancy all this is dealt with in a later section as will CE marking.

In summary, the following diagram Figure 7.2 shows the above three step process.

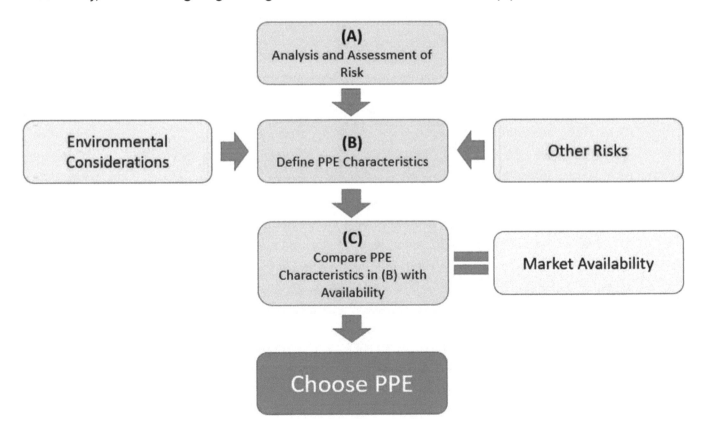

Figure 7.2 PPE assessment flowchart

The following table is reproduced from Article 5 of the 89/656/EEC Use of Personal Protective Equipment Directive. This will aid the aid in the identification of all hazards.

		RISKS																	
		PHYSICAL										CHEMICAL						BIOLOGICAL	
		MECHANICAL					THERMAL		ELECTRICAL	RADIATION		NOISE	AEROSOLS			LIQUIDS		GASES, VAPOURS	HARMFUL BACTERIA	HARMFUL VIRUSES	MYOCOTIC FUNGI	NON MICROBE BIOLOGICAL ANTIGENS
PARTS OF THE BODY		FALS FROM HEIGHT	BLOWS CUTS IMPACT CRUSHING	STABS CUTS GRAZES	VIBRATION	SLIPPING FALLING OVER	HEAT FIRE	COLD		NON IONISING	IONISING		DUST FIBRES	FUMES	VAPOURS	IMMERSION	SPLASHES	GASES VAPOURS				
HEAD	CRANIUM																					
	EARS																					
	EYES																					
	RESPIRATORY TRACT																					
	FACE																					
	WHOLE HEAD																					
UPPER LIMBS	HANDS																					
	ARMS																					
LOWER LIMBS	FOOT																					
	LEGS																					
VARIOUS	SKIN																					
	TRUNK ABDOMEN																					
	PARENTAREL PASSAGES																					
	WHOLE BODY																					

Figure 7.2b Risks table

Personal protective equipment is, in principle, intended for personal use although the PPE directive gives the possibility for it to be worn by more than one person by ensuring that this does not create any health or hygiene problem. I would very much doubt if that would be acceptable nowadays.

In addition, the employer must consult the workforce every step of the way. I cannot stress how important that this is when choosing arc flash protection and when carried out correctly, it has a positive effect on morale and eradicates misuse and non-compliance. There is a good example of how this can be done in Chapter 11: Electrical Utilities. It describes the journey taken to equip the entire workforce at Northern Powergrid with new FR uniforms and where worker involvement was a study in how it can be done well. Notwithstanding the case for worker involvement from a sensible management perspective, the "Framework Directive" (European Council Directory 89/391/EEC (EU Workplace Health and Safety Directive)) states that workers shall be informed of all measures to be taken with regard to the health and safety of workers when PPE is used by workers at work. The "Framework Directive" also mandates that there shall be consultation of workers and participation by them in all matters relating to the provision of PPE. As a minimum, the employer must first inform the worker of the risks against which the wearing of the personal protective equipment protects him. The employer shall then arrange for training and shall, if appropriate, organize demonstrations in the wearing of personal protective equipment.

7.3 CE Marking

Employers should ensure that any PPE they buy bears a 'CE' mark and complies with Regulation (EU) 2016/425 on personal protective equipment. (The full title is Regulation (EU) 2016/425 of the European Parliament and of the Council of 9 March 2016 on Personal Protective Equipment, repealing Council Directive 89/686/EEC). Not to be confused with the Use of Personal Protective Equipment Directive (89/656/EEC) covered previously, the regulation lays down requirements for the design and manufacture of PPE, which is to be made available on the market, in order to ensure protection of the health and safety of users and establish rules on the free movement of PPE in the European Union. It requires manufacturers to CE mark their products to show compliance. If you use PPE for providing protection against arc flash hazards, you should ask for confirmation, from the supplier, that the PPE certified satisfies the requirements of the PPE Directive.

Following Brexit, things are different in the United Kingdom. The UKCA (UK Conformity Assessed) marking is a new UK product marking that is used for goods being placed on the market in Great Britain (England, Wales and Scotland). It covers most goods which previously required the CE marking. The UKCA mark will not be recognised for products being placed on the EU market.

The UKCA marking alone cannot however, be used for goods placed on the Northern Ireland market, which require the CE marking or UKNI marking. There is a separate Northern Ireland Protocol which, for as long as it is in force, Northern Ireland will align with all relevant EU rules relating to the placing on the market of manufactured goods.

7.4 Types of Arc Flash PPE

When considering protecting persons against the arc flash hazard, much of the focus seems to have been on providing clothing made from FR fabrics in the form of shirts, polo tops, trousers coveralls and so on. However, as I mentioned earlier, it has been shown that the parts of the body which have been most frequently affected by arc flash have been the hands, arms and face. Some years ago, when researching the subject, I remember being told by a HSE Specialist Electrical Inspector about an horrific injury to a young man injured in an arc flash accident. He suffered permanent and debilitating injuries to his respiratory system due to the inhalation of toxic and super-heated gases. I have been very fortunate to have made the acquaintance of experts in hand and face protection with whom I have used this observation on several occasions. Fortunately, the standards for face and hand protection have caught up in recent years and details of the different types of arc flash protection and related standards are detailed as follows.

7.4.1 FR Protective Clothing

Compliance is required with Standard IEC (EN) 61482-2: 2018, Live working - Protective clothing against the thermal hazards of an electric arc - Part 2: Requirements. This standard will give some assurance that the garments comply with rigorous minimum standards of inspection and testing. The subjects covered are; tear resistance, tensile strength, burst strength, shrinkage and other effects of cleaning, arc thermal performance, marking/labelling, user instructions specification and flame resistance. There are also inspections of construction and workmanship, size designation and ergonomics, ageing and threads and closures.

All flame resistant garments are made from fabrics which are in turn, made from fibres which are woven together to form the fabric. The term "Inherently Flame Resistant" means that the fibres are naturally flame resistant and when woven into a fabric require no further treatment and can be made into garments to withstand levels of the thermal hazard from an arc flash event. The flame resistance is an inherent property of the polymer chemistry and will not diminish during the lifetime of the garment.

Then there are those garments made from "Permanently Treated Fabrics". They tend to be made from cotton fibres which acquire their flame resistance from a treatment received after it is woven into fabrics. Permanently treated fabrics were often characterised by the treatment washing out over a period of time which is determined by the frequency of washes that the fabric was exposed to. This is less of a problem nowadays although the life expectancy of inherently flame resistant fabric appears to be longer through anecdotal evidence and published data. Those manufacturing FR garments from permanently treated fabrics would say that they are more likely to naturally wear out before losing flame resistance.

Whether inherent or treated, sewing thread used in the construction of FR garments have to be inherently FR fibre except for where threads in seams that have no influence on protection, such as hems and pocket seams.

My own preference is for garments made from inherently flame resistant fibres and my reason is supported by my house building analogy. A house may be built out of several different materials. I love to see a house made out of kiln fired bricks, but I have to accept that a house made out of concrete blocks will probably be cheaper and perform just as well. The fact is, that the cost of the bricks or the concrete blocks, will be a fraction of the total building costs and the manufacturer of the materials will not be the company that builds the house. This is the same for PPE where the weavers will mostly be separated from the manufacturers of the fibres. So, for larger bespoke procurement of PPE it is preferable, in my view, to specify the type and even the producer of the fibre and drive the value for money through the weaving and manufacturing of the garments. For smaller, off the shelf procurement exercises, the comparison of apples with pears starts with being aware of the performance of the various fibres that go into the garments. The following information will help with the comparison.

7.4.2 User acceptability

Comfort. If the garment is to be worn as part of everyday workwear, then it needs to be comfortable. In addition to the thermal performance per weight of fabric, it has to feel comfortable as well as be lightweight. This is often termed as the hand of a fabric and is assessed by the reaction obtained from your sense of touch. The challenge with measuring fabric hand is that everyone perceives things differently. The air-permeability of the fabric and a good garment design are parameters, which determine the comfort of protective clothing. Comfort is not only another key aspect for acceptance of the clothing by workers, but if the garments are very uncomfortable then this may even increase the risk of an electrical accident or even trip, slips and falls.

Colour. In addition to being comfortable, if the garment is to be worn as part of everyday workwear, then it needs to look good and colour could be important to gain user acceptability. The problem is that inherently flame resistant fibres may be limited in colour ranges so this is where compromises may be required. Durable appearance and colour fastness are also a key element for acceptance of clothing by workers and evidence should be available from the supplier.

Life Expectancy

There needs to be a discussion about the life expectancy of any garment and which factors will affect it. From the data that I have seen, it is likely that inherently flame resistant fabrics will last longer than permanently treated fabrics, but purveyors of the latter type may be able to give a contrary viewpoint. The life cycle cost could vary widely and an inability to provide evidence could result in a much greater expense for an inferior product.

Laundering

The standard allows for both washing and dry cleaning, but it is down to the manufacture to make clear how the garments have to be cleaned in their instructions. This could have a significant effect on life cycle cost. The manufacturer may state a maximum number of cleaning cycles, in which case the requirements for limited flame spread have to be met after the stated cleaning cycles. Otherwise, if no maximum number of cleaning cycles is specified, the flame test has to be carried out after five cleaning cycles.

Underwear

Arc rated under garments are available and their provision does ensure that workers are equipped with non-flammable clothing throughout. Another advantage is that higher levels of protection can be obtained when used as part of a layered solution as described above. In Europe however, the provision of undergarments as part of a PPE system has some drawbacks. The employer must provide it free of charge and they also have an audit inspection role to ensure that it is being worn. The complications such an audit will bring, I will just leave to your imagination. From a practical perspective, it may be better to mandate that the worker must not wear undergarments or clothing made of materials that would add to the hazard of possible burns. This is where the specifier may find that the open arc Incident Energy Limit (ELIM) rating has a part to play, insomuch that there is a zero chance of burn injury. It follows that undergarments cannot ignite for properly rated FR PPE in an arc flash incident. See section 7.5.2 for further information about ELIM.

Labels

Every arc resistant garment has to be labelled which should show the manufacturer, type of product, size, and care requirements. In addition, the garment will have the IEC arc flash symbol shown in Figure 7.3 which denotes protection against the thermal effect of the electric arc. It should also show the level of arc thermal protection in the form of an arc rating ELIM and/or arc protection class (APC 1 or APC 2). In addition to ELIM the lower value of either ATPV or EBT (Energy Breakopen Threshold) can be also be added. All basic essential guidance with respect to cleaning should be given by labels or other marking of an item of PPE. More detailed additional information will be given on instruction for use/manufacturer's leaflets.

Figure 7.3
IEC Garment Symbol

Layering

The layering of arc rated garments can have a beneficial effect on the final arc rating as well as comfort. For instance, if an arc rated jacket of 8 cal/cm² is worn over an arc rated shirt of 8 cal/cm² it would suggest that the final Arc Thermal Protection Value (ATPV) will be of the order of 16 cal/cm². The fact is that the final ATPV if the garments were to be used together would be considerably higher than that. This is often used by suppliers to produce higher comfort levels from higher rated garment systems. It should be stressed that layering has to be tested as a system together to assure their protection level. The following diagram (Figure 7.4) shows how two layers of 9.4 cal/cm² results in a final tested ATPV of 33 cal/cm². See section 7.5.2 for further information about ATPV.

FIGURE 7.4 LAYERING EXAMPLE

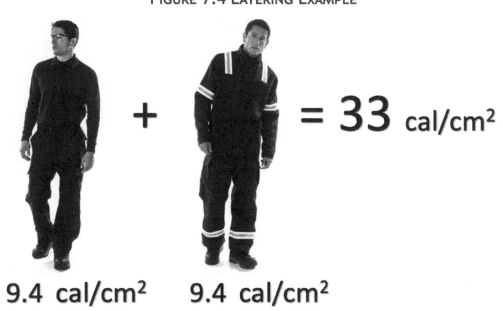

Figure 7.4 Layering of Protection

7.4.3 Use and Maintenance of PPE

Personal protective equipment may be used only for the purposes specified, except in specific and exceptional circumstances. It must be used in accordance with instructions and the instructions should be available and understandable to the workers. It is the obligation of the manufacturer to give clear instructions about use, care and maintenance of the PPE which should be followed up by the employer and workers. The instructions should specify in particular the cleaning and drying methods and means. They should also detail the storage and inspection regime to be adopted. All basic essential guidance with respect to cleaning should be given on the label or other marking of an item of PPE. More detailed additional information shall be given on Instruction for use/manufacturer's instructions/ user instructions.

When choosing PPE, it is at the bottom of the hierarchy of risk control measures. As such, all lower order risk control measures need to have a much greater level of monitoring and review. The goals of the monitoring and review process should be:

- That the PPE is being worn and inspected correctly.
- Periodic review of the hazard/risk assessment may need a revision of the use of PPE.
- That the inspection, storage, cleaning and decontamination is being carried out in accordance with manufacturer's instructions.
- That the PPE is maintained in a safe, usable condition to provide the intended protection to the user.

7.4.4 Training

The usual training, assessment and authorisation of persons is necessary and appropriate criteria must be applied, see Chapter 6: Process, Policies and Procedures. Workers need to be trained on how to use their PPE correctly, prior to the PPE being introduced into service. As a minimum, the training should include:

- Information concerning limitations and capabilities of the PPE.
- How the PPE works and what the PPE will and will not protect.
- How to follow the risk assessment of which PPE is a part.
- Issues of sensory deprivation and how these issues can be mitigated.
- How to read and correctly interpret the information which is given on label or other instructions.
- How to use, wear and inspect the PPE.
- How to store the PPE when not in use.
- Information concerning arrangements for handling, cleaning and decontamination.
- How to determine when the PPE is no longer fit for purpose.
- How to obtain replacements.
- The dangers of using PPE which is contaminated by inflammable liquids or substances.

It is recommended that a holistic approach be adopted for training the users of PPE and that they be engaged in the process of the provision and use of protective measures right from inception. Simply providing written instructions or information may not be effective and practical demonstrations and formal training will lead to better acceptance.

7.4.5 Record keeping

The keeping of records will assist in the management of the arc flash PPE. A full life history can be built for each item, from manufacture to disposal. The record keeping will allow the duty holder to understand the life cycle of the PPE and help with monitoring and review. The life cycle costs can be better understood when the cost of maintenance and durability are built in. This will allow for improvements in future decision making in respect of replacement and maintenance.

7.4.6 Routine examination

PPE should be examined preferably by the user before and after use. The overall risk assessment should detail the examination which should also be to match the thermal protective performance to the item of PPE. The PPE should also be formally inspected when the item has been cleaned and records kept about condition. Anyone undertaking inspections should be appropriately trained.

7.4.7 Cleaning and Ageing

Cleaning should be strictly in accordance with the manufacturer information including care instructions. Based on this information, the duty holder should determine the arrangements for care and provide a process for the cleaning and decontamination of arc protective PPE. This also gives an opportunity for the formal examinations and recording of condition.

Professional or industrial cleaning is the favoured method depending upon the severity of use and allowing home cleaning should only be done under strict considerations.

Ageing can be effectively forecasted by the manufacturer by indicating the maximum number of cleaning procedures. Deterioration due to ageing has an effect on the performance of the arc protective PPE and can be accelerated by exposure to chemicals and other agents, physical exposure such as radiation and heavy wear and tear.

7.5 PPE Test Methods for the Electric Arc Hazard

In Europe, for protective clothing, there are two tests that are used to determine the level of protection against the thermal effects of electrical arcs. The two tests are referred to colloquially as the "Open Arc Test" and the "Box Test". Their full reference is given below.

Open Arc Test - IEC 61482-1-1:2019, Live working – Protective clothing against the thermal hazards of an electric arc – Part 1-1: Test methods – Method 1: Determination of the arc rating (ELIM, ATPV and/or EBT) of clothing materials and of protective clothing using an open arc.

Box Test - IEC 61482-1-2:2014, Live working - Protective clothing against the thermal hazards of an electric arc - Part 1-2: Test methods - Method 2: Determination of arc protection class of material and clothing by using a constrained and directed arc (box test)

Both the box test and the open arc test are entirely different in their methodology and application. Furthermore, it is not easy to determine equivalence between the output of both methods and the risk assessment process uses a different approach in each case. Depending on the needs, either or both standards can be specified.

The most commonly used output from the open arc test is an arc thermal performance value (ATPV) which is a numerical value of incident energy attributed to a product (material or equipment) that describes its properties of attenuating the thermal effect of energy generated by an open arc. For example, if the ATPV is 12 cal/cm^2, then that particular specimen or item of PPE is capable of attenuating an incident energy of that same level or less to a "safer" value of 1.2 cal/cm^2 for the wearer. Therefore, anyone undertaking a numerical risk assessment of incident energy level on site, can directly compare the results obtained to the thermal withstand properties of available PPE.

The main output from the box test is an Arc Protection Class. The box test method defines two arc thermal protection classes APC 1 and APC 2. The two classes cannot be linked to numerical risk assessment of incident energy level on site in the same way as the open arc method and do not provide an arc thermal performance value (ATPV).

7.5.1 Open Arc Test Setup

The open arc test set up uses two stainless steel electrodes at a distance of 30mm to 300mm using a decoupled source giving a prospective short circuit current between 1 kA and 20 kA. The test voltage needs to be sufficient to sustain an electric arc across the electrodes which is typically between 400 V and 700 V. It is the duration of the arc that is used to vary the effective energy on the surface of the specimen to be tested or the protective clothing which can be typically between 200ms and 2000ms.

In the case of PPE, the specimen garments or samples of material are placed around the arc electrodes spaced at 120 degrees at a distance of 300 mm. Calorimeters are placed behind the materials to be tested to measure the energy let through the sample. At the same distance, further calorimeters are placed that are exposed to the electric arcs. These calorimeters directly exposed to the electric arcs measure the actual effective energy to which the garments or sample materials are being subjected.

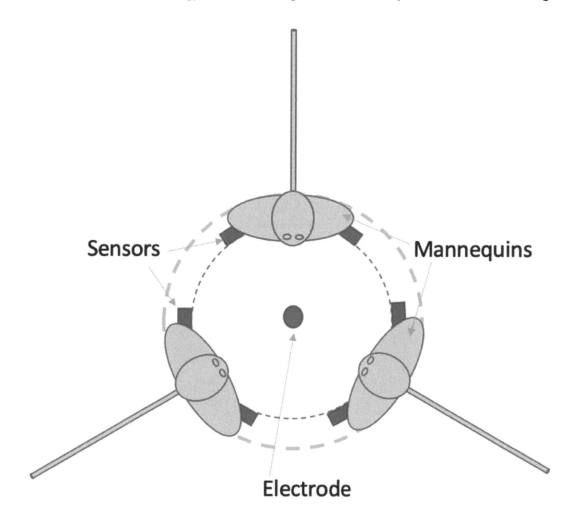

Figure 7.5 Open Arc Test Setup

The above diagram shows the mounting of three mannequins at 120 degrees around the central electrode. These could be materials or a mixture of mannequins and materials. The additional calorimeters are placed as shown at a predetermined distance from the arc and the sternum is held centrally to the arc.

7.5.2 Open Arc Output Values

The Arc Thermal Performance Value (ATPV), as mentioned previously, is the main output for the open arc testing that is required to match PPE to the site calculated incident energy levels. In order to determine the ATPV, the electric arc current will be 8 kA, the distance between the electrodes 300mm and the distance from the arc 300 mm. A calculation of ATPV is made from the measured calorimetric values of the directly exposed calorimeters and the calorimeters covered by material samples. Tests are performed using different arcing durations to give the resultant variable energy exposure.

I mentioned a value of 1.2 cal/cm^2 as a "safer" level and this has been derived from work which was carried out by biophysicist Alice Stoll and Maria Chianta in the 1950s who from their experiments in the US Navy, created the Stoll Curve. They carried out experiments on volunteers by subjecting them to heat on their forearms and recorded the temperature and time at which the individuals responded to the pain. This was performed across a range of radiant heat fluxes, and the burns that developed blisters after 24 hours were recorded as second-degree burns. The Stoll Curve is used with a graph of heat flux against exposure time. The point at which the heat flux cross the Stoll Curve is identified as the point at which a human would feel pain and be at risk of second-degree burns. The ATPV gives the incident energy of a material that results in a 50% probability that sufficient heat transfer through the specimen to cause the onset of second-degree burn injury based on the Stoll Curve. In other words, half the points measured would be above the curve and half below.

Incident Energy Limit (ELIM) is the numerical value of incident energy attributed to a product (material or equipment), below which the values of all product responses are below the Stoll curve and without breakopen. Put simply, the ELIM differs from the ATPV in that there is 0% chance of a second-degree burn rather than a 50% probability. Where this is applied to a garment, I understand that this additional safety factor will usually result in a derating of about 10%. It is sometimes regarded as a purely academic measure. This is because the ATPV would have to be exactly equal to the actual arc flash to create the conditions of a 50% chance of a minor burn. In my practical experience of thousands of arc flash calculations in the field, I cannot recall a single occasion when these conditions have existed. If used as part of a risk assessment, it will be usually an additional safety factor on top of how the incident energy calculations are applied.

Breakopen Threshold Energy (EBT) is the numerical value of incident energy attributed to product (material or equipment) that describes its breakopen properties when exposed to heat energy generated by an electric open arc test. It represents the highest incident energy exposure value on a fabric where the garments do not exhibit breakopen and is the value of incident energy at which breakopen occurs with 50% probability. A breakopen is defined as a minimum of a 1.6 cm^2 hole formation. Various fibre types act in different ways but each one can breakopen before the burn prediction level is reached. Workers are assumed safe if the arc rating of their clothing (or ATPV value) exceeds the electric arc incident energy as calculated in the worst-case scenario of a risk assessment.

7.5.3 Open Arc Test – Further Discussion

PPE tested to the open arc method can be linked directly to quantitative risk assessments on site and calculations through the IEEE 1584 Guide for Performing Arc Flash Calculations. The calculations will provide incident energy and an arc flash boundary and, although discussed elsewhere in this guide, some additional care is required due to the interpretation of the outputs from the open arc test according to IEC 61482-1-1.

Exposure to heat flux of 1.2 cal/cm² for the duration of 1 second, i.e., exposure to incident energy of 1.2 cal/cm², can produce the onset of second-degree burn or partial thickness burn of the bare skin. Limiting the incident energy exposure at the skin surface to no more than 1.2cal/cm² means that you can still receive some burn injury, however the primary objective of arc flash protection has been to minimise the injury and probability of death. In general, if the prospective incident energy exposure at a given location is below 1.2cal/cm², no additional thermal protection is required for the worker. This is achieved by the determination of a boundary beyond which the incident energy is less than 1.2cal/cm², or through shielding or application of properly designed specialist PPE capable of withstanding the thermal effects of the arc. Where protection against the thermal effects becomes necessary it must be emphasised that PPE does not prevent the accident happening in the first place.

Personal Protective Equipment (PPE) used for arc flash protection includes garments made from FR fabric. When tested to the open arc test IEC 61482-1-1, the fabric is designed to provide a thermal barrier and limit the incident energy exposure at the skin surface to no greater than 1.2 cal/cm². Although FR fabric may burn when exposed to a flame, it is designed to stop burning when the flame is removed. It also must not break or burn open and expose the skin directly to the flame. As discussed previously, the ATPV represents the incident thermal energy that results in a 50% probability that sufficient heat transfer through the clothing is predicted to cause the onset of a second-degree burn injury, or - more colloquially - the ATPV is the incident thermal energy that the clothing can support before the wearer would suffer second-degree burns.

As described above, when carrying out open arc testing according to IEC 61482-1-1, an ELIM is obtained which similar to the ATPV in that it is derived by statistical means from the test data but is slightly more conservative. Whereas the ATPV is based on the 50% probability of crossing the Stoll curve, the ELIM is based on no points crossing the Stoll curve. The reason that the ELIM was introduced was because of disagreement in Europe over the concept of a 50% chance of a burn, albeit a minor one. This was based upon the European PPE Directive. So, the ELIM was introduced which made sure that there would be a zero chance of a burn where the ELIM rating of the garment is greater than the prospective incident energy level. An advantage of using ELIM instead of ATPV is that the prescribing of specific undergarments is less of an issue. See the earlier section on undergarments. In other words, if properly prescribed, there is zero probability of the ignition of garments worn beneath ELIM rated PPE.

It might be worth pointing out that having spoken with various people in the industry, the ATPV model is valid and there are no injuries reported where the ATPV was greater than or equal to the actual incident energy level in an arc flash accident.

The introduction of ELIM creates a dilemma for the duty holder when determining the arc flash protection and how to match the site calculated data. For instance, the IEEE 1584 Guide for Performing Arc Flash Calculations will produce an arc flash boundary which is the distance from a prospective arc source at which the incident energy is calculated to be 1.2 cal/cm². This is the boundary, outside which, no FR PPE is required according to US and Canadian consensus standards. In addition, working on any equipment where the incident energy level was calculated to be less than 1.2 cal/cm^2 at the working distance would not require FR PPE. Theoretically, there could be a 50% chance of a burn at both these distances where the incident energy level was 1.2cal/cm^2. These are the various options that the duty holder could adopt to accommodate ELIM principles.

- **Options for the arc flash boundary (AFB)** – Either stick to the 1.2 cal/cm^2 limit for the arc flash boundary or build in an additional safety factor on top of the IEEE 1584 calculations. For instance, if the arc flash boundary was 1 metre add 10% safety factor making the AFB = 1.1 metre. A better solution would be for a site wide standard arc flash boundary to be adopted. Say if the arc flash boundary varies from between 0.5 and 1.8 metres, then round it up to 2 metres. It is easier to apply and easier to understand.
- **Options for incident energy levels requiring no PPE** – Either stick to the requirement of 1.2 cal/cm^2 above which FR PPE would be required or reduce this incident energy level at the working distance by a factor of approximately 10%. For instance, FR PPE will be required in all case where the incident energy level at the working distance was above 1.08 cal/cm^2. A better solution would be that FR PPE should be worn for any interactions with energised equipment.

7.5.4 Further Requirements for Protective Clothing

When tested according to the open arc test, the protective clothing made of the tested material has to have an arc rating as described below in order to meet with the requirements of IEC 61482-2:2018 - Live working protective clothing against the thermal hazards of an electric arc.

Arc Rating is the numerical value attributed to a product, that describes its protective performance when tested in accordance with the open arc test. The arc rating can be the arc thermal performance value (ATPV), the breakopen threshold energy (EBT) or the incident energy limit (ELIM). A manufacturer may assign a lower arc rating value to a material or protective clothing than the value resulting from testing.

Protective clothing needs to meet a minimum arc thermal protection value, where the ELIM is at least 130 kJ/m^2 (3.2 cal/cm^2). In addition, it can have a minimum arc thermal protection, where the lower value of ATPV and EBT is at least 167 kJ/m^2 (4 cal/cm^2). In case only either ATPV or EBT can be determined, this value will be at least 167 kJ/m2 (4 cal/cm^2).

Due to the limitations of test apparatus at very high energy arcs, no arc rating above 4186 kJ/m^2 (100 cal/cm^2) shall be assigned to garments.

7.5.5 Box Test Setup

The box test is set up as a directed arc from a plaster box and consists of a vertical arrangement of a pair of electrodes between which the arc is ignited. See Figure 7.6.

The open circuit voltage is 400 V (50 Hz) and the arc gap is fixed at 30mm, with an aluminium upper electrode and a copper lower electrode. The duration of the arc is also fixed for a period of 500ms. The arc emission is focused outwards and corresponds to extreme effects under the relevant conditions. The test takes account not only of the directed effect of heat, but also the effect of hot metal particles and splashes that accompany actual electric arc faults. The box test will test a sample of fabric up to a

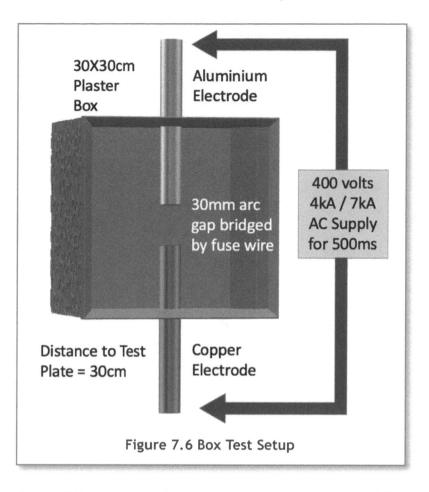

Figure 7.6 Box Test Setup

maximum of five samples and then a specimen of the garment. It is a very simple test not requiring decoupled power supplies and the only variability is the prospective short circuit current which is either 4 kA or 7 kA. The result is either pass or fail resulting in either thermal protection classes Arc Protection Class (APC) 1 (at 4kA) or Class 2 (at 7kA).

Figure 7.7 The Author observing the box test at Ilmenau University in Germany

The materials to be evaluated are fixed onto a plate at a horizontal distance of 300mm from the electrodes. There are two calorimeters are set in this plate which determine the heat flow densities and are used to judge whether the wearer of clothing made from this material would have suffered second-degree burns. A further requirement is that there is no afterburn lasting for more than 5 seconds and that holes are no bigger than 5mm in any direction. Attention is paid to the total behaviour of the clothing, whether for example the clothing can be opened easily after testing and that fasteners are still functional.

7.5.6 Box Test Output Values

To comply with IEC 61482-2:2018 Live working — Protective clothing against the thermal hazards of an electric arc a material and the clothes made from it must at least pass Class 1 of the "Box Test" in order for clothing from this material to be termed "clothing for protection against thermal effects of electric arcing". (Alternatively, the material must at least have an ATPV of 4 cal/cm^2 and the clothing made from it must pass the "Open Arc Test" as described earlier)

7.5.7 Arc protection classes

When tested according to Box Test, the protective clothing made of the tested material is assigned one of two arc protection classes (APC) This will either be APC 1 is subjected to a prospective short circuit current of 4 kA for 500ms or APC 2 at a prospective short circuit current of 7 kA for 500ms. Protective clothing has to demonstrate a minimum arc thermal protection of APC 1 in order to be given an arc rating according to the standard. A fabric will pass the test:

- If the heat transferred behind the fabric cannot cause a second-degree burn. (In box testing, no crossings of the Stoll Curve are permitted)
- If the fabric passes a visual assessment covering:
    - No after-flame time is below 5 seconds.
    - No melting to the inner side of the fabric.
    - No holes larger than 5mm.

Future visual assessment requirements for head, face and eye protection will probably not allow breakopen, dripping or molten matter inside the protector.

NOTE: The necessary arc protection class is determined by risk analysis. Guidance for the appropriate selection of the arc protection class is provided in other separate guidelines, such as the ISSA Guide and DGUV-I 203-078, Thermal hazards for electric fault arcs – Guide to the selection of personal protective equipment for electric work. There is a calculator which is available through this guide on ea-guide.com. In addition, some further assistance in predicting arc severity for Box Test PPE can be found in Chapter 14: Hazard and Severity Calculations. My opinion is that APC 1 or 2 ratings may be the basis for the selection of PPE, as long as the actual expected exposure situation can be considered to be less severe than the specific exposure condition simulated during the box testing according to IEC 61482-1-2.

7.6 Summary of Minimum Arc Thermal Performance for Protective Clothing from IEC 61482 -2

In summary the following diagram, Figure 7.8, shows the minimum requirements that protective clothing will need in order to meet EN 61482 -2.

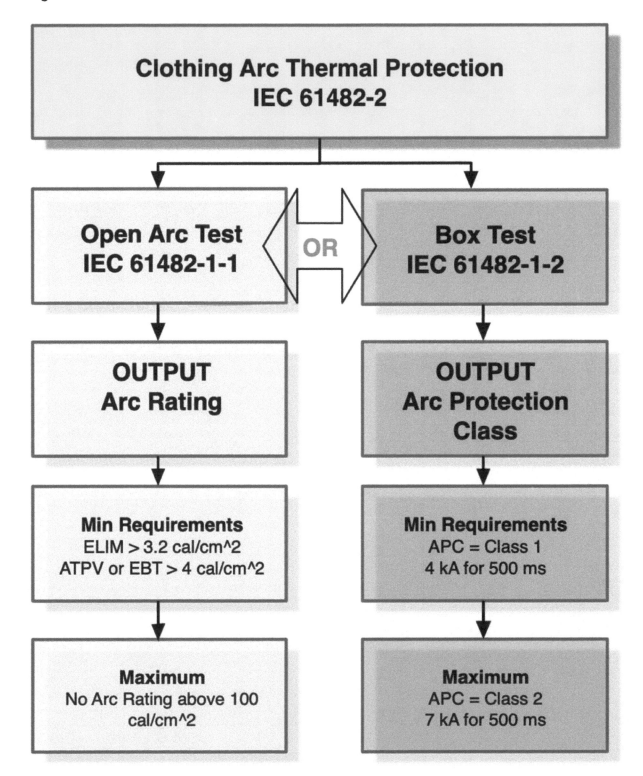

Figure 7.8

7.6.1 Box Test versus Open Arc

There has been considerable argument over the relative advantages and disadvantages between the box test and open out test methods. This has even manifested itself in comments made in International standards and official guidance on the subject of protection against the thermal effects of electrical arcs. The following are examples of such comments.

The following comment is a quotation from the National Foreword to the box test standard – EN 61482-1-2:2007 from the UK technical committee PEL/78. "It is the opinion of UK technical committee PEL/78 that the "box test" does not provide the user with a realistic and reliable test. A premise of this test is that the fault currents will not exceed 4000A or 7000A and the worker will not be closer than the specified distance from the arc (in reality this cannot be guaranteed)". "The UK technical committee PEL/78 believes that the open arc test IEC 61482-1-1 will provide the best way to determine whether a particular material will provide the best protection for the worker for any given job".

The following comment came from the *DGUV 5188 Thermal Hazards from Electric Fault Arc -Guide to The Selection of Personal Protective Equipment for Electrical Work. "On the other hand, Arc-Man test results lead to the so- called Arc Thermal Performance Value, or ATPV. In this context, the incident energy is determined according to a statistical methodology, by which a 50% probability exists of suffering second-degree skin burns behind the PPE. Even if an electric fault accident is relatively improbable, the EU directive regarding PPE allows no interpretation of PPE that would tolerate such injury. For this reason, as a matter of principle, such test methods should not be used within the EU". This is very clear but as I have said previously, to be wearing PPE which had an ATPV at the exact level of the actual arcing incident energy in a real-life situation would be improbable. It is also the case that there are safety factors built into how the incident energy calculations are applied. This is where the new test parameter the Incident Energy Limit (ELIM) has been introduced. ELIM differs from the ATPV in that there is 0% chance of a second-degree burn rather than a 50% probability.

*Note - DGUV is the German Social Accident Insurance which is the national, compulsory program that insures workers for injuries or illness incurred through their employment.

Both methods are not perfect because how could they be? Electrical arcs are violent and often unpredictable events, so the research has served well to provide two methods of meeting international standards. I have therefore provided information on both standards and also calculator tools for determining the ATPV/ELIM associated with the open arc method as well as the (APC) box test. In this case you have the tools to do some comparisons. The following are my thoughts on the relative merits and drawbacks to each method.

The setup of the box test closely resembles low voltage utility termination equipment or cut out, the similarities being copper and aluminium electrodes, the voltage and the prospective current. The closeness of the sample to the arc is also plausible for this type of equipment which is often installed in enclosed areas. Lots of low voltage cut outs are still mounted in cupboards under stairs for instance and service cupboards are notoriously tight on space. Indeed, "the (box test) conditions are intended to represent the practical situation in low-voltage installations and low-voltage networks. The qualitative results (re-burning, melting, hole formation etc), derived at either 4 kA/500ms or at 7 kA/500ms, may be used to assess whether any given test system (fabric or protective clothing) would be appropriate under exactly the same real working conditions". (ISSA Guideline for the selection of personal protective clothing when exposed to the thermal effects of an electric arc) However, these conditions are dissimilar to conditions outside electrical utilities. For instance, very few modern industrial installations have a mixture of copper and aluminium and in most situations the working distances are greater. In addition, the fault currents within industrial facilities are much greater so the box test is unlike "the same real working conditions."

The box test is much harder to correlate to quantitative risk evaluation via arc flash calculations such as IEEE 1584. In addition, the arcing current and time to clear are two variables that are essential to obtain an accurate estimate of arc power/incident energy. Perhaps one could use IEEE 1584 and then apply that to obtain a disconnection time to determine the box test APC.

The electrical source for the open arc test is a large generator setup which has to be decoupled from the mains supply. The cost of these facilities is high which explains the reason why there are so few of them in the world. The electrical source for the box test is a directly coupled mains transformer making it much easier and cheaper to perform the test.

The open arc test is carried out on the garment whereas the open arc test only needs the fabric. In addition, there are only 4 exposures required for the box test making it cheaper. The drawback is that the box test does not subject the whole garment to the thermal effects of the arc and therefore, the open arc test could be seen as giving a truer prediction of the garment performance in an actual arc event.

Advocates of the box test would point out that the open arc method can only capture the radiated thermal effects of an arc whereas the box test specimens will be more effected by splashes of vaporised and molten materials.

The box test has, as previously explained, got just two exposure levels which are a prospective short circuit current of 4 kA for 500ms (APC1) and 7 kA for 500ms (APC2) The material must pass either to be given an arc protection class (APC) as well as fulfil the qualitative criteria in the standard. Having just two levels means exposures higher than APC2 cannot be evaluated which could mean that exposures above 12 to 20 cal/cm^2 cannot be protected. APC1 can return an approximate ATPV of 4-8 cal/cm^2 and Class 2 is from 12-20 cal/cm^2. However, it can be argued that pass/fail makes interpretation more straightforward as no statistical analysis is required to determine classification levels. The downside of a pass/fail result is that repeatability for fabrics that only just pass is not as easy to identify compared with specific values given for ATPV or ELIM.

I have seen evidence of where a major utility company who were using box test garments in the knowledge that the supplied PPE may be underrated for certain site conditions. This was based upon what the perceived level of acceptance to protection would be and the phrases used were "comfort/ risk balance" and "some protection was better than none at all". I am sure that a better strategy would be to do the field studies first and then determine the PPE requirements in terms of ATPV based upon the task, distance, fault levels, protection arrangement as well as the prevention techniques detailed in Chapter 5: Prevention.

My recommendation would be for the box test rated PPE to be used for utility type low voltage systems only for instance to replicate potential hazards in; service entrance boxes, cable distribution, street lighting cut-outs, cabinets, distribution substations or comparable installations and/or where the arc is directed to the front of a worker at the height of the breastbone. For all HV and LV industrial and commercial situations, utilities, power stations and transport infrastructure I would strongly recommend the use of open arc rated PPE.

7.7 Head, Face and Eyes Protection

At present, there is no IEC standard for arc protection for the head, face and eyes and IEC 61482-2 is only for garments. There is, however, a standard being written with the probable title of "Live Working – Eye, Face and Head Protectors against the Effects of Electrical Arc – Performance Requirements and Test Methods".

In the meantime, there are a choice of products available that may give the protection that you need. The advice is that you must use a face shield with a reputable arc rating. Where fabric forms part of a head assembly as shown in the following image Figure 7.9, then the fabric can be tested to either the box test or the open arc test and given either an arc protection class (APC) or an arc rating. There are also many products on the market that are tested to ASTM standards (American Society for Testing and Materials) such as the ASTM F2178 / F2178M - 20 Standard Specification for Arc Rated Eye or Face Protective Products. If it tested to that ASTM standard, then an ATPV is available but not an ELIM. There are German standards available that require testing to box test to BG-Prüfzert GS-ET-29:2010-02: Supplementary requirements for the testing and certification of face shields for electrical works and suppliers will often quote both an ATPV as well as a box test APC 1 or 2. A face shield which is tested to American standards will be valid as they are similar to the IEC open arc test.

The face shields are often made from energy absorbing formulations that can provide higher levels of protection from the radiant energy. The shields are tinted so advice should be sought from the manufacturer in respect of light transmission class, as they may compromise colour perception visual acuity. This is obviously an important issue and supplementary illumination of the work area might be necessary when these types of arc protective face shields are used.

There may be a requirement for the protectors to cover not only eyes and face, but also further the entire head, including the neck. This needs to be addressed in the risk assessment. If necessary, this can be provided by a hood to cover all the areas simultaneously.

The images below (Figure 7.9) show typical face shields mounted with a fabric chin and neck guard on a helmet or a brow guard and often come with an open arc test ATPV of around 12 cal/cm² and typically box test APC2. These examples can be supplemented by an FR balaclava to give 360-degree protection.

Figure 7.9 Face Protection Images Courtesy of J K Ross Ltd

Figure 7.9

7.8 Hand Protection

Hand protection comprising rubber insulating gloves and leather over protectors have been worn for electric shock protection for many years. The good news is that the same combination provides fairly good protection against arc flash as well. There are, however, arc flash protection gloves that have been tested to both box and open arc test methods and they are readily available in Europe. Furthermore, products are available with high level of cut resistance and chemical protection. Care must be exercised if choosing leather over protectors to ensure that they are unlined or lined with non-flammable, non-melting fabrics. Heavy-duty leather gloves with minimum thickness of 0.7mm meeting this requirement have been shown to have ATPV values in excess of 10 cal/cm². (Source - NFPA 70E:2015 Standard for Electrical Safety in the Workplace) The specialist arc flash protection gloves are available with an ATPV of > 50 cal/cm² and/or ARC 1 or 2 classification.

The standard for insulating gloves is IEC 60903 Electrical Insulating Gloves. There are insulating gloves on the market in Europe that claim that over protectors are not required although care is always required that the dielectric properties are not compromised through heavy use. From personal experience, leather or specialist over protectors do not reduce dexterity if sized properly and paradoxically can sometimes improve the grip of small components. To determine the size, measure the distance around the palm of the hand, just behind the knuckles and choose the next available size up.

7.9 Foot Protection

I am not aware of any standards but normal industrial heavy-duty leather footwear provides a degree of arc flash protection. You may think that the feet will be the body part furthest away from an arc flash incident and usually you would be right. There are situations however, when they are the closest to a possible arc such as phasing out a pavement mounted link box for an electrical utility for instance.

7.10 Hearing

Although hearing protection for arc flash protection is much debated, there are no specific standards requiring arc rated hearing protection. Future standards may deal with the need for testing FR ratings if part of an overall head and face protector.

The general consensus of opinion appears to be that standard canal inserts to reduce sound pressure exposure above 100 dB will be adequate. In this way the inserts (and the person's ears) will be protected by a shield, hood or balaclava. Canal inserts typically give a SNR (Single Number Rating) of 30 and most large suppliers have SNR 33 canal inserts which will give the desired protection. Canal caps are not favoured they tend to be held in place with plastic that can melt and tend to be less efficient.

7.11 Pragmatic Approaches to PPE Selection

At this stage, if you have followed the 4P approach, any PPE that is prescribed will be as a last resort and for the residual risk only. Where an ATPV is used, it is likely that the majority of calculated incident energy levels at control panels and switchgear will now be below 1.2 cal/cm², or the threshold of a second-degree burn. Furthermore, the vast majority of incident energy levels in industrial and commercial environments will be in single figures. The remainder may be quite high in magnitude. This fact will allow the duty holder to think about protection in terms of categories based upon the risk assessments on a local basis. So, if the majority of levels are below 8 cal/cm² for instance, then a simple approach will be to have a standardised level of 8 cal/cm². At this level, there are many options available giving a very comfortable single layer solution. This has been called "basic protection" and every situation above this level as "enhanced protection". (reference Jim Phillips – Arc Flash Hazard Calculation Studies) For those enhanced protection situations there may be a simple solution involving layering which is

Learning Points

➢ PPE can only be prescribed in Europe after the employer has analysed and assessed the risks which cannot be avoided by other means.

➢ Follow the assessment of PPE which shall involve an analysis and assessment of risks which cannot be avoided by other means.

➢ Involve the users at an early stage to get acceptability and compliance.

➢ Try to follow a simplified pragmatic approach to PPE Selection.

Chapter 8

Data Collection

Data collection is, without doubt, the most important part of the arc flash risk assessment process. It involves skill, experience and patience as well as a high level of competence. Even if information such as single line diagrams and electrical protection information is already in existence, data collection can take as much as 50% of the effort. I call this the discovery phase as it often feels as though you are a bit of a detective trying to unearth missing pieces of evidence. This chapter is dedicated to the practicalities of collecting data and whilst the following will cover the scope of a complex study of an average commercial/industrial electrical network, the basic principles of acquiring data for a one-off risk assessment still apply.

It is vital that the planning and preparation phase is undertaken with some rigor if the overall data collection process is to be successful. The following paragraphs and the flowchart in Figure 8.1 suggest that there is a good deal of work to do before going to site.

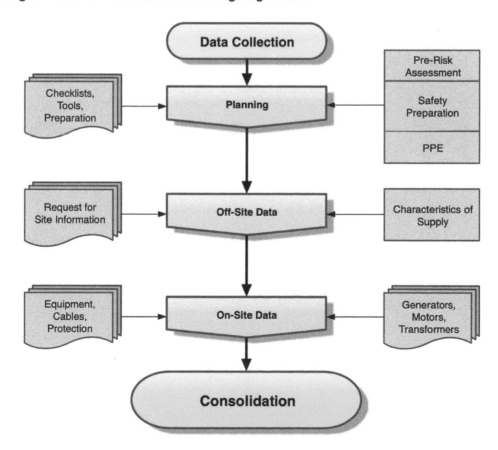

Figure 8.1 Data collection flowchart

8.1 Planning

Whether you are employed directly by the site owner or a contractor, it is a good idea to set time aside for the planning phase and this is the first step in the above flowchart. If you are the duty holder or you are not in charge of the site, then it is a good idea to get a check list of things to do and documents drawn up and dispatched to the person who is. The following items are suggestions for the checklist.

8.1.1 Check list

⇒ Project plan and set a date for a pre-start meeting.
⇒ Get the supply source data – this needs to be one of the first actions which is to request the characteristics of supply from the local utility. More details plus a sample letter follow later in this chapter.
⇒ Submit the draft risk assessment and method statement. This can be tabled and completed at the pre-start meeting.
⇒ Details of the company safety inductions.
⇒ Establish the nature of the work as non-intrusive survey and inspection.
⇒ Information on site hazards and substances to be provided by the client.
⇒ Access and/or authorisation approvals required in particular to switch rooms and substations.
⇒ Rules for access to inspections at height.
⇒ Site safety rules including electrical safety rules or guidance.
⇒ Seek permission for any photography that may be required.
⇒ Establish who will be available at the start meeting and throughout the project and their authorisation level.
⇒ Establish levels of accompaniment
⇒ Access to switchgear, shut down periods.
⇒ Establish a base on site – quiet area or office with internet connection.
⇒ Make a list of everything that you need to complete the study. Some examples are below.
  o Valid single line diagrams.
  o Accurate site plans.
  o Design data including cable sizes, impedances and lengths.
  o Previous short circuit, protection coordination or arc flash reports.
  o Installation and commissioning certificates.
  o Records of electrical installation test and inspections.
  o Maintenance records.
  o Details of electrical switchgear embargoes or known defects.
  o Schedule of protection settings including author.
  o List of assets or assets register.
  o Schedule of motors and types of starting/speed control.

8.1.2 Tools for the Job

It is a good idea to get organised for the survey and the following is a non-exhaustive list of tools for the job.

⇒ Accurate callipers for measuring the outside diameter of cables.
⇒ Cable sizing tables.
⇒ High resolution camera.
⇒ Non conducting tape measure and/or laser measure.
⇒ Measuring wheel.
⇒ Non-metallic torch.
⇒ PPE – see separate section on planning for safety.
⇒ Quick calculators to determine "what ifs" for such as electrode configurations and dynamic safety risk assessments. (Provided with this guide)
⇒ Calculators for sense checking relay settings as you go along. (Provided with this guide)
⇒ Hazard report forms – to report serious hazards that are spotted to the duty holder.

Anything that is taken onto site must be declared to the duty holder along with test certificates for all equipment including tools and access equipment. Any electrical equipment should be tested and calibrated where necessary. Permission may be required for cameras as some sites are commercially or security sensitive.

8.1.3 Planning for Safety

The risk assessment and method statement should have been completed and agreed and most importantly followed by all stakeholders. I have resisted the urge to provide a sample document here because of the plethora of different environments that may be encountered. The following considerations are, however, the salient points that need to be addressed.

Risk of inadvertent switchgear tripping. What is the consequence of inadvertent switchgear trips? Consider who may be harmed and how, maybe the inspection work is in a hospital setting or supplies are to emergency services or for environmental protection purposes.

Authorisation levels. Make sure that you are competent for the tasks which also means that you can recognise at all times the limits to your own competence. This is a discovery mission, and it may be unfamiliar equipment or circumstances arise in which case it is time to stop. This is particularly the case with high voltage switchgear and the HV authorised person for the site must be in attendance when entering and inspecting HV substations and enclosures.

Accompaniment. Avoid lone working when carrying out inspections and entering switch rooms and substations. Anyone giving accompaniment should know what to do in an emergency so it will usually be necessary to be accompanied by someone who is competent to make the system safe and avoid injury.

Opening of equipment cubicles and doors to facilitate inspections. This should always be with client approval and adherence to electrical rules and procedures.

Interaction with switchgear should be avoided. The operation of switchgear by an external service provider should not be permitted. Any electrical measurements that may be necessary should be subject to a separate risk assessment and in collaboration with the duty holder.

Personal Protective Equipment – The provision of PPE is essential for any inspection where there is a risk of injury from all hazards including the arc flash hazard. The requirements for the selection and use of arc rated PPE have been comprehensively covered earlier in this guide, not forgetting other hazards including electric shock. So, what level of arc rated PPE should you wear?

From a practical point of view, a strategy of wearing a single layer of comfortable garments, complemented with appropriate head, face and hand protectors will probably work for the majority of low voltage inspections up to 400 volts alternating current. The next question is, which inspections are in the majority and which ones could be in the minority where this strategy will not work?

This is where the calculation tools that are available with this guide at ea-guide.com will be of assistance. Armed with the tools, technical experience and site knowledge, it is now possible to approach the viability of the single layer solution by exception.

Upstream Protective Device Less Than 100amperes.

For equipment, which is supplied by circuit breakers and fuses below 100 amperes, the vast majority of incident levels will be below 1.2 cal/cm^2 at a working distance of 600mm regardless of the prospective short circuit current. This approach will work where the upstream device is a current limiting fuse such as NH DIN or BS88 fuses or for miniature circuit breakers up to IEC 60898-1 types B&C or moulded case circuit breakers where the instantaneous setting is no greater than 10 times the nominal current rating. This is assuming that the protective device is capable of withstanding and breaking the prospective short circuit current.

At very low prospective short circuit current levels, the time to disconnect may increase and as a result, so will the incident energy level. There is a calculator which accompanies this guide that lists LV Fuses and MCBs up to 125A calculator against which the actual incident can be checked against. It is recommended that this is used as it includes all five IEEE 1584 Electrode configurations.

Upstream Protective Device More Than 100 amperes.

Use of the generic circuit breaker calculator, available with this guide (at ea-guide.com), will allow you to build up a list of exceptions where the incident energy levels are below the arc rating of the PPE for the protective devices greater than 100 amperes.

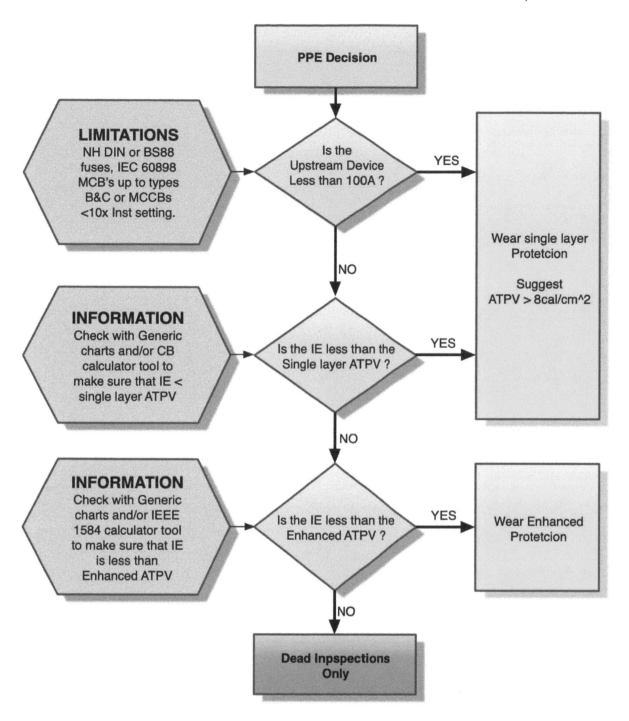

Figure 8.2

The working distance can be added to this and with care there should be no reasons to be any closer than 600mm for inspection work. The usual working distance is 450mm for live work at 400 volts so, this increase in distance will result in a reduction of incident energy level of about a third. (This is given for illustrative purposes only and will require validation for the particular site conditions) For the remainder of inspections, the calculators will allow a quick estimation of the actual incident energy level. By this time, the prospective short circuit current will be known, making this process easier to adopt. For all other instances, do not take risks, and always arrange for a shutdown if necessary. The flow chart in figure 8.2 should assist in arriving at the correct arc rating for the task of inspection.

Finally, do not forget the other risks such as electrical shock. If there is any risk of contact with live parts, which I would hope would be extremely small, then it may be necessary to provide insulating gloves and mats.

Working at height. If access is required for inspections at height, then this will need to be risk assessed. Whilst there may or may not be rules for access equipment onsite, the use of client's own equipment needs careful consideration. This will always be a collaborative approach and the operation of hoists or scissor lifts by untrained personnel must be avoided. I mentioned the addition of a high-resolution camera to the "Tools for the Job" earlier and a zoom lens has allowed me to read equipment labels from the ground. For example, I have been able to get information from nameplate data from high level busbar trunking systems without having to arrange for access equipment.

8.2 Supply Source Data

8.2.1 Source Short Circuit Current Data and the Characteristics of Supply

Collecting data from the electricity supplier or third-party supplier in respect of the characteristics of supply can sometimes be seen to be extremely challenging particularly in respect of obtaining a realistic estimate of the available fault current. This is, however, the starting point for any arc flash risk assessment whether it be for a whole complex network supplied at high voltage or for an individual low voltage switchboard. Best practice dictates that we can obtain the required information by enquiry, calculation or measurement. There may be a case to adopt all three methods depending upon the complexity of the project, but the overriding consideration must always be confidence in output from any risk assessment.

8.2.2 Enquiry

Where we need to ask for the characteristics of supply this is usually from a public electricity company or maybe a third-party supplier.

Fundamental facts that you should know about the information from an electricity supplier.

1. There is sometimes an obligation on electricity supply companies to provide the maximum prospective short circuit current. This is very specific in most supply agreements and often a legal duty. The problem is that the definition of the maximum prospective short circuit current, is often the theoretical design capability of the connecting cables and switchgear and is unlikely to be the fault level that appears at a connected customer's supply terminals.

2. The electricity supplier does hold all the information that you will need to undertake an arc flash risk assessment. In other words, as well as knowing what the maximum prospective short circuit current is, they will know what the actual or realistic prospective short circuit current is. The tricky bit is being able to speak to a person within the organisation who is knowledgeable on the subject.

3. If the electricity supplier changes the characteristics of supply, then this can affect the output from any calculations.

4. There is a fundamental shared objective for electricity supply companies to cooperate across operational boundaries in respect of safety. A supply company can do things that can affect the safety of connected companies and connected companies can do things that can affect the safety of supply company operatives.

From my experience, it can sometimes take time and effort to get the required information and so it is recommended that the very first thing that is put in place is an enquiry letter. The following is a sample enquiry letter.

You will notice that I have deliberately kept the information as simple as I can and by not overcomplicating the message with additional non-essential information. Once contact has been made with a technical engineer who can help, more nuanced information can be sought if necessary. Experience has shown me that once contact with the right person has been made, engineers are usually very helpful particularly when they are speaking to fellow engineers.

Sample Enquiry Letter

To - Electricity Company & Address

Dear Sirs/Madam,

Electricity Safety Risk Assessment – Request for Supply Characteristics Information

Client name

Site Address and intake substation reference number (or customer reference number)

We are currently carrying out an arc flash risk assessment and request your cooperation in order to complete the work. This work is essential to develop strategies to protect our personnel and your own whilst on site against the dangerous effects of arcing faults on our equipment. The work will involve a protection grading and an arc flash hazard study. In order to perform the calculations, we will require fault level and protection co-ordination information at the point of connection to our site. The information requested is as follows.

1. Maximum and minimum symmetrical prospective short circuit fault current at the supply terminals.
2. X/R values at maximum and minimum fault current.
3. Maximum and minimum earth fault current levels.
4. The type, rating and settings of the protective device or devices nearest to the supply terminals.

Please be aware that the study will require the minimum fault currents as well as the maximum as lower fault currents can create the highest level of hazard in some circumstances. I will be pleased to discuss our requirements in greater detail and my contact details are as follows.

Name, address email and telephone number.

Thank you in anticipation of your cooperation and I look forward to hearing from you in due course.

Yours faithfully etc...

The letter should go out on the company letterhead for whom the supply contract exists and not on letterheads from third party contractors or consultants. An alternative is to get an accompanying letter of authorisation signed from the company from whom the supply contract exists if you are acting on their behalf.

You will also notice that I have not included the units in the enquiry. It is likely that different supply companies will present their data in different ways. For instance, the information may come back as an R+jX impedance or MVA or kA plus X/R ratio. The main thing is not to place barriers in the way of the response.

8.2.3 Calculation

Once the source prospective short circuit current is known then calculations can be made to determine the short circuit current at the equipment. One of the calculators which is appended to this guide will facilitate this. Given the source high voltage short circuit current and X/R ratio on the transformer primary, PSCC can easily be calculated anywhere in the low voltage network.

8.2.4 Measurement

Another way of determining the prospective short circuit current at source, or anywhere in the electrical distribution system, is by measurement. Most earth loop impedance testers give us the ability to return a short circuit current value, but extreme care must be taken in the interpretation of results. Such instruments become more inaccurate the closer to the source transformer because of the higher levels of inductance. There are, however, high precision loop testers available that will give an ohmic value of resistance and reactance as well as the balanced three phase short circuit current. Accuracies of 1 milliohm are possible from such an instrument.

That sounds great so why do we not just measure the PSCC everywhere instead of by enquiry and calculation? Well, the answer is that you could, but I would exercise some caution because of the following.

1. Live testing is classified as live working insomuch that there will be interactions with live conductors. As we have learnt in this guide, live work must pass the test of reasonableness before being sanctioned.
2. For high precision loop testers, the test current is likely to be in the order of three hundred amperes rather than the usual 25 amperes that one would find with standard earth loop impedance testers. Whilst the instruments that I have used are extremely robust and well capable of handling such test currents, they do require another degree of care in application.
3. Contributions to PSCC under fault conditions such as from induction motors cannot be picked up using an instrument and may need to be factored in. The IEEE 1584 incident energy level

calculator which is appended to this guide does allow for an estimate of motor contribution to be calculated in addition to the source PSCC.

In conclusion, an understanding of the prospective short circuit current, which is available at source, is a particularly important early step. What I mean by the word understanding is, what are the realistic maximum and minimum values likely to be? And what factors are significant to the outcome of the arc flash incident energy calculation? Bearing in mind that the IEEE 1584 calculations have built in confidence factors such as the normal and lower arc current calculation, the exercising of care in determining the source impedance will result in greater credibility and accuracy of the overall risk assessment. I would therefore suggest that instead of making a choice between enquiry, calculation, or measurement, that a combination of the three factors will lead to validation and confidence of the characteristics of the supply sources.

8.3 The Survey

Make friends with the guys at the facility and, if not done at the initial meeting, spare the first hour or so on the first day to carry out a presentation to them. Tell them what you are doing, how you are going about it and exactly why you are doing the work. This is a time to allay some fears that they may have in respect of over engineered and cumbersome PPE and following the process in this guide will lead to sensible management of the arc flash hazard. In this way you can win them over and obtain much more cooperation and hopefully assistance in the information gathering process.

8.3.1 Existing information

It is important to understand that the information that is gathered in the discovery phase will have a bearing on the accuracy of the final calculations. There is often a plethora of information available, some new and some old and often conflicting. Very often circuits have been installed that have not been recorded in the proper way and there will be no information available. There is usually a single line diagram for the high voltage system available at each site and if not, that would signal a major non-conformance. The main thing to understand, is that any information, which is made available from records on site, must always be questioned and validated.

8.3.2 Collecting data

My approach has been to break down the whole of the electrical distribution system into small chunks and work through circuit by circuit. You will eventually have a valid single line diagram with detailed circuit data, but the raw data can be populated onto a spreadsheet which is split into four separate sections with each row representing one circuit. The first section can be about the panel or equipment detailing the equipment reference, location details, busbar geometry, type and where it is fed from. The second section is all about the protection for the outgoing circuit with information including circuit reference, make, type, rating, CT ratio, breaking capacity, and settings. The third section relates to the outgoing circuit which will have the cable details such as cable reference, type, size, cores per phase, length and method of installation. The fourth section can pick up the load type and included here will be large motors and particularly induction motors that can contribute to prospective short circuit current. Each row of data for every circuit should be populated in full and any missing data will be easily recognised from the grid. The grid should be large enough to be able to view a whole panel or switchboard onto one single sheet. Keep it simple and try to avoid duplication and in that way, nothing will be missed. Sketch out any circuits that may be more than discrete radial circuits such as bus sectioning and rings.

In addition, take high resolution photographs which are catalogued to the above panel/circuit references. This will serve as an aide memoire in the validation process.

Data collection software is available, usually from the manufacturers of complex modelling software. The programs work well with the associated programme but are not easily interchangeable. See Chapter 13: Complex Software Guide.

8.3.3 Equipment Data

8.3.3.1 Electrode Configuration

If using the IEEE 1584 Guide for Performing Arc Flash Hazard Calculation 2018, it will be necessary to determine the electrode configuration at the equipment that is being modelled. Electrode configuration is the term used for the orientation and arrangement of the electrodes used in the testing performed for the IEEE 1584 model development. These are detailed in the IEEE 1584 guide and there is some further guidance to be had within annexe C and annexe G2 of said guide. The following information is presented to further assist the engineer to determine which configuration is necessary when applying the IEEE 1584 model in a European context.

There are five different electrode configurations used in the laboratory conditions to create the formulae and these are as follows.

- VCB: Vertical conductors/electrodes inside a metal box/enclosure.
- VCBB: Vertical conductors/electrodes terminated in an insulating barrier inside a metal box/ enclosure.
- HCB: Horizontal conductors/electrodes inside a metal box/enclosure.
- VOA: Vertical conductors/electrodes in open air.
- HOA: Horizontal conductors/electrodes in open air.

The task is to establish which electrode configuration to use as the choice will determine not only the arcing current but also the incident energy & arc flash boundaries. If there are some doubts about which configuration is the best fit for the actual configurations on site, then the quick calculators provided with this guide will allow some "what if" type considerations to work out which is the best fit in terms of worst-case scenarios. This may be necessary because there may be multiple configurations in one piece of equipment.

IEEE 1584: 2018 Electrode Configurations

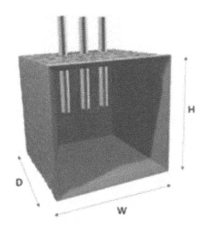

VCB: Vertical conductors/electrodes inside a metal box/enclosure

The VCB configuration uses three equally spaced vertical conductors in a metal box enclosure. The characteristics of this configuration is that the arc is driven away from the source of the supply and downwards to the bottom of the enclosure and out towards the worker. The setup is very similar to the 2002 edition of IEEE 1584.

VCBB: Vertical conductors/electrodes terminated in an insulating barrier inside a metal box/enclosure

The next configuration is VCBB which stands for vertical conductors or electrodes in a metal box enclosure that are terminated into an insulating barrier. What has been demonstrated is that the arc hits the barrier and then the plasma cloud is driven outwards and towards the worker.

HCB: Horizontal conductors/electrodes inside a metal box/enclosure

This is HCB where horizontal conductors or electrodes are pointing outwards inside a metal enclosure. This configuration can be particularly aggressive as the plasma from the arc is directed from the ends of the electrodes and directly at the worker.

VOA: Vertical conductors/electrodes in open air

The final two configurations are in open air. The VOA configuration on the left shows vertical conductors or electrodes in open air. This was also in the 2002 edition of IEEE 1584.

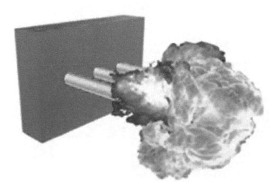

HOA: Horizontal conductors/electrodes in open air

The HOA configurations is similar to VOA but is with horizontal electrodes in open air.

8.3.3.2 Examples of Electrode Configurations

Configuration	Description	Enclosure	Examples
VCB	Vertical conductors	Metal	Metal Cut Outs, Industrial Service Units, factory-built switchboard assemblies, LV isolators and switch fuse units.
VCBB	Vertical conductors into insulating barrier	Metal	Busbar trunking systems, tap off units, factory-built switchboard assemblies
HCB	Horizontal conductors	Metal	Transformer LV cable boxes, Pavement mounted link boxes, Busbar trunking systems, tap off units.
VOA	Vertical conductors	Open Air	Terminations onto overhead lines, Live LV Jointing, Pole mounted transformer terminations.
HOA	Horizontal conductors	Open Air	Open transformer pad terminations, overhead line spans, Live LV Jointing, Open LV feeder pillars.

8.3.3.3 Conductor Gap

There are various schools of thought on just how far you have to go in arriving at an exact figure when collecting data on conductor gap. The IEEE 1584 Guide for Performing Arc Flash Hazard Calculations has a number of default gaps for determining the possible arc distance between conductors. The latest version (2018) suggests 152mm for 15kV switchgear and MCCs, 104mm for 5 kV switchgear and MCCs, 32mm for low voltage MCCs and panel boards and 13mm for cable junction boxes. There was a survey some years ago by Jim Phillips on brainfiller.com and he asked if people use the standard default gaps from IEEE 1584 rather than measure them on site. It was almost unanimous that people do use the default gaps. The reasons given were that it was implausible that individuals were actually going to measure the equipment and anyway there are multiple gaps for any given piece of equipment, so which one do you use?

The fact is, that the conductor gap does make a difference otherwise the calculations would not allow for it. I am afraid this is where engineering judgement comes in once again. Looking at any given piece of equipment, you may not know the exact arc gap, but you will have a good idea of what the arc gap is not. In other words, an estimate to see if the default gaps from IEEE 1584 are reasonable and failing that err on the side of a larger arc gap. At lower voltages a larger gap will tend to give a reduced arcing current and the most incident energy per unit time. The tools that have been provided with this guide allow you to play with the conductor gap figure and make up your own mind on a case-by-case basis.

8.3.4 Equipment Condition

It is strongly recommended that there is an assessment of the condition of the equipment during any survey to assess arc flash risk. The following is a good template for assessing this risk which will feature in the final report.

One of the criticisms of US standards in respect of electrical safety and arc flash was that there was a perceived link directly from incident energy levels into the wearing of PPE. So, regardless of what condition the equipment was in, or whatever the task was, if the calculated incident energy level was 40 cal/cm^2 then you had to wear a 40 cal/cm^2 suit. At one extreme, workers were as a result, switching perfectly safe brand-new switchgear which was often internal arc protected. In 2012, I was asked to peer review a paper titles Risk Management of Electrical Hazards by the author Daniel Roberts, which was presented to the IEEE Electrical Safety Workshop in Florida and introduced the concept of "as low as reasonably practicable" (ALARP) principles based upon internationally accepted OHS Risk Management. The paper was well received as was my paper that I presented to the same conference alongside co-author Jim Phillips on "A European View of Arc Flash Hazards and Electrical Safety". In the years since then the US consensus standard NFPA 70E Standard for Electrical Safety in the Workplace has developed more of a risk-based approach to the provision of PPE.

Part of the assessment of risk from NFPA 70E has introduced the concept of normal operations and providing that the equipment meets the six conditions below. Normal operation of electric equipment shall be permitted where all of the following conditions are satisfied:

1. The equipment is properly installed and commissioned. In other words that the equipment is installed in accordance with applicable industry codes and standards and the manufacturer's recommendations.
2. The equipment is properly maintained. Which means that the equipment has been maintained in accordance with the manufacturer's recommendations (plus company engineering instructions) and applicable industry codes and standards.
3. The equipment is used in accordance with manufacturer's instructions.
4. The equipment doors are closed and secured.
5. All equipment covers are in place and secured.

6. There is no evidence of impending failure. Which means that there is evidence such as arcing, overheating, loose or bound equipment parts, visible damage, or deterioration.

This gives a template for estimating the likelihood of an occurrence of an arc flash and if all the conditions are met then the likelihood is very low. Should any of the conditions fail then there is a greater likelihood of an arc flash and therefore other measures including use of PPE as a last resort would be necessary.

8.4 Protection Data

When collecting data on protective devices, the vital question must be, which device will operate in the event of an arcing fault. A good example of this is where there is an incoming circuit breaker as part of a factory build switchboard or a distribution switchboard. If there was to be a fault on the load side of this circuit breaker, can it be assumed that this device will clear the fault without being involved in the arcing event? In other words, can the arc propagate to the line side of the circuit breaker? If it can, then the next upstream device must be considered as the means of disconnecting the arcing fault in which case the time to trip will be longer leading to greater severity. A similar course of action must be considered if there are doubts that the incoming circuit breaker is capable of disconnecting the arcing fault because of a lack of maintenance or obsolescence.

The collection of protection data will be aided by some preparation in respect of existing records and also building up a library of protective devices and associated curves. Mainstream manufacturers have improved access to time current characteristic curves, and many provide the means to print the curves and settings. There are also tools available with this guide which will assist in the modelling of curves and settings to help validate data. I have given some description on the relay types as some younger engineers may not be familiar with the history and background on the older technologies.

8.4.1 Current transformers

If good records exist, then this will make the collection of data so much easier. Be aware that current transformers (CTs) may have been changed from initial installation for all manner of reasons but most commonly for protection upgrades. The CT ratios will often be stamped inside the small wiring panel of the switchgear or on the rear of the switchgear for feeder protection. See Figure 8.3.

Figure 8.3 Current transformer nameplate

There will also be many occasions where the CT ratios are not marked on the switchgear. What is the breaker feeding? A 400-ampere feeder is likely to have CT primary rated at 400 amperes and 10/11kV transformers up to 1000 kVA are likely to have CT primary rated at 100 amperes or below. Dual wound current transformers are very common, so it is very important to ascertain which winding is used. Further detective work may reveal the CT ratios on metering where these are not separately supplied. The use of a clip-on ammeter to the small wiring can also be compared to metered values for the primary circuit if fitted. If the feeders are at high voltage, then remember the authorisation/competence levels as spelled out previously under the earlier section on "Planning for Safety".

Always adopt a verification process and sense check what the records say or the markings on the switchgear. For instance, maintenance test, commissioning records or asset registers will usually indicate the CT ratios. If these records are not available because of neglect, then this is a good indicator that the equipment will be unreliable in a fault situation. This should be a prompt to have the necessary maintenance carried out at the earliest opportunity.

8.4.2 Electromechanical Inverse Definite Minimum Time Relays

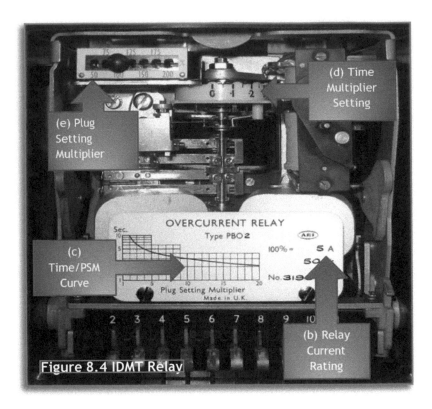

Figure 8.4 IDMT Relay

(d) Time Multiplier Setting

(e) Plug Setting Multiplier

(c) Time/PSM Curve

(b) Relay Current Rating

Induction disc electromechanical relays like the overcurrent model in Figure 8.4 have been around for over a hundred years giving reliable service to the protection of electrical systems. Even when electronic static relays had proved to be reliable and accurate, electromechanical relays carried on being specified particularly with electrical power companies well into the 1980s. As a result, they are commonly found in high voltage equipment and occasionally on low voltage switchgear to this day. The principle of operation is that electromagnets that are fed circuit current, induce current into a moving metal disc creating torque and when great enough, causes the disc to rotate. The disc is retained by a spring and if the magnetic force is great enough, rotation will occur and will operate contacts which in turn will trip the circuit breaker. The disc will rotate at a speed which is commensurate with the magnitude of the current in the measured circuit. The time dial, shown on Figure 8.4 adjusts the distance that the disc has to rotate to close the contacts. The slowest will require almost a full revolution of the disc. This creates the time multiplier setting (TMS) the range of which is usually between 1.0 to 0.1, and the time to operate at the 0.1 setting is reduced to one tenth of that at the 1.0 setting.

The electromagnetic primary winding has a number of tappings that are brought out to the plug setting bridge on the front of the relay, see Figure 8.5.

Dependent upon which plug is used, this will result in the plug setting multiplier (PSM). The plug setting will determine the current at which the relay will start to operate. The tapping range is usually 50%, 75%, 100% (For a 5-ampere relay, 100% means 5A), 125%, 150%, 175% and 200%. Some manufacturers may use 1.5, 3.5, 5, 6.25, 7.5, 8.75 and 10A. For example, for

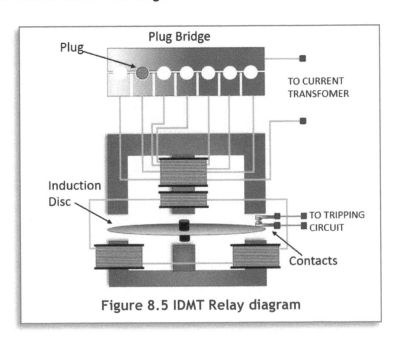

Figure 8.5 IDMT Relay diagram

a CT rating of 100/5A, if the relay is set to operate at 5A then the plug setting will be = Relay current setting/5A = 5A/5A = 100%. For a relay to operate at 2.5A, the plug setting (for this example) will be 2.5A/5A = 50%.

So, what is the operation time of an electromechanical relay for a given arcing current? To do this you will need the following information.

a) Ratio of current transformers
b) The current rating of the relay.
c) The time / PSM curve of the relay.
d) The time multiplier setting.
e) The plug setting multiplier position.

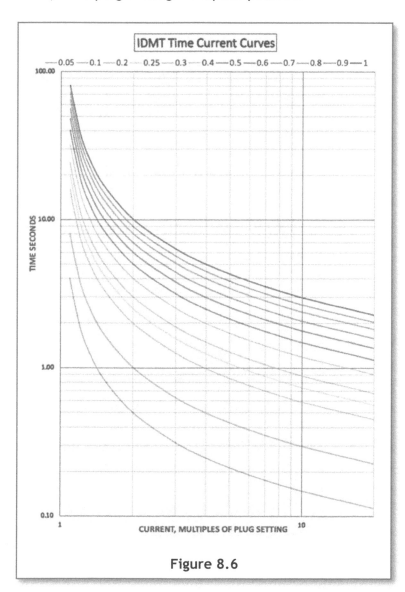

Figure 8.6

The current transformer (CT) ratio (a) is covered in an earlier section, just make sure that the current rating of the relay (b) matches the secondary of the CT. The relay current rating can usually be found on the front of the relay cover as shown on the previous page in Figure 8.4. The time/PSM curve (c) is usually provided on the relay cover as well as in manufacturer's literature. If the source of the information is the latter, it may be presented as the characteristic curve shown in Figure 8.6. The time multiplier setting (d) and the plug setting multiplier position (e) can be taken directly from the relay, as shown in Figure 8.4.

As an example, if we have the following information, we can determine the relay time to operate; CT ratio = 400/5, current rating of the relay = 5A, the curve is shown in Figure 8.4 above, the time multiplier setting TMS = 0.2 and the plug setting multiplier position is 100%. We have calculated an arcing fault current of I_{arc} =5000 amperes.

The relay current = I_{arc} x CT Ratio = 5000 x 5/400 = 62.5 amperes.

The plug setting multiplier is 100% meaning that the current multiple = Relay current/rating

= 100% x 62.5/5 = 12.5 times

Using the curve for a TMS of 0.2, the corresponding time to trip is therefore 0.57 seconds.

Bear in mind that there is the time for the circuit breaker to operate to take into account as well as relay inaccuracies. Both these factors need to be added to the relay trip time before applying this to calculate the incident energy. Some further information to follow.

Warning - Never touch the moving parts of electromagnetic IDMT relays. Actions such as in trying to manually turn the disc or bending contacts can warp moving parts or introduce acidity which may result in decay to contact surfaces.

8.4.3 Static Relays

Static Relays were first introduced the 1960s and gained in popularity in the 1970s. They were the next generation on from induction disc electromechanical types. So called static relays meant that they have no moving mechanical parts and monitored the voltage and current incoming waveforms by analogue circuits.

They were an electronic replacement for electromechanical relays designed to carry out the same functions but performed better as they are more accurate, are fast acting and caused less burden on current transformers. Knowledge of electromechanical IDMT relays meant that settings and, in this case, data collection is fairly straightforward. The static relay shown in Figure 8.7 was a very popular model which arrow switches to select and set the required time current characteristic shown on the front panel. This same format was adopted by several manufacturers. For data collection, this makes the process fairly straightforward as all the information is on display. Most user manuals are easy to obtain by google even if some of the relays are over thirty years old. Manufacturers are keen to make information accessible, even about

Figure 8.7 Static Relay

obsolete products so no need to pay specialist companies for photocopied documents.

8.4.4 Standard IDMT Overcurrent Characteristics

The standard IDMT overcurrent characteristics are expressed as a mathematical formula according to the IEC 60255-3/BS142 standards and are often the basis of the electromechanical time current curves. The fact that the curves are based on a mathematical formula makes it easy to sense check the relay settings and of course electronic relays use the same curves.

The equation for operating time from IEC 60255 trip curves is as follows.

$$t(I) = TMS \left\{ \frac{k}{(\frac{I}{Is})^{\alpha} - 1} \right\}$$

Where, Is = the current setting and I = the actual current and k and α are the curve type constants. See table below.

There are four curve types used in IEC 60255 (BS 142) which are: standard inverse, very inverse, extremely inverse and long-time standard inverse. For electromechanical relays, the curves are fixed for the particular relay model. So, in other words, if a standard inverse function is required then that is built into the specification of the relay. It will be necessary therefore to check the relay model with manufacturer's data to determine the curve type.

Curve Type	Constant	
	k	α
Standard Inverse	0.14	0.02
Very Inverse	13.5	1
Extremely Inverse	80	2
Long Time Standard Inverse	120	1

Note that US relay characteristics to ANSI and IEEE follow a similar format to the above equation but have a further constant L which is added to the expression.

The tools that are available with this guide (through ea-guide.com) provide curves that are based upon all the above IEC 60255 curve types with inputs for current transformer ratios.

8.4.5 Digital Protection Relays

Digital Relays were introduced in the 1980s and instead of analogue circuits, microprocessors were used to implement relay functions. Digital relays use analogue to digital convertor of all measured analogue quantities and use a microprocessor to implement the protection function given previously. Within 10 years both the electromechanical and the solid-state static relays had been mostly replaced by the digital protection relay in industrial applications, although this has proceeded more slowly the electrical utility distribution systems. The perceived complexity of digital relays and a need for simplicity in utility distribution systems may be an explanation of this phenomena. There was also a great deal of nervousness about the millennium bug prior to the year 2000 which no doubt had an effect. (Google the millennium bug to find out what the fuss was all about.)

That apart, digital relays have advantages over electromechanical and the solid-state static relays such as a much higher level of functionality, integration and additional monitoring functions. They also provide flexibility of functions meaning that a single unit can provide many more protective functions other than overcurrent and earth fault. They are generally more accurate, are self-checking and able to communicate with other digital equipment.

8.4.6 Numerical Protection Relays

The next generation on from static/digital relays are the numerical multifunction protection relays which also perform control and monitoring functions. These devices combine the functions of protection relay, data logger and multi-functional display in a single unit creating cost-effective power system management as well as protection. Communications, self-testing, remote operations, energy management and multiple protection functions can be performed in a single unit.

Data collection can be tricky sometimes as there are often password-controlled operations meaning that levels of access can be limited. The possession of a technical manual (usually a few hundred pages) is absolutely necessary as there is nothing generic about the units in the way that they are set or interrogated. That said, the protection, particularly for overcurrent and earth fault follows the same curves and simulates the mathematically generated operating characteristic of earlier generations.

8.4.7 Time Fuse Links or Time Lag Fuses

Time fuse links are extremely popular particularly for distribution transformer protection particularly in public utility companies for this type of application. This is a very cost-effective means of protection requiring no tripping supply and does create an efficient and a reliable means of disconnecting transformers under fault conditions. The characteristics are easily obtained from the manufacturer or technical standard number which is found on the barrel of the fuse link as seen in figure 8.8.

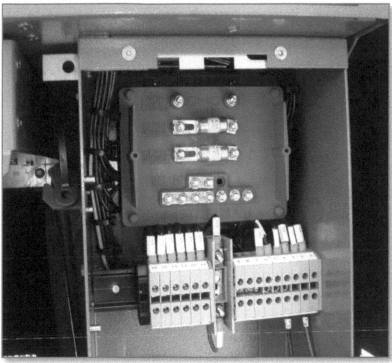

Figure 8.8 TIME FUSE LINKS

Operation of time fuse links can be explained with the aid of the wiring diagram shown in Figure 8.9. As can be seen, the circuit consists of 3 dual ratio CTs, 2 Direct Acting Trip (DAT) solenoid type overcurrent coils connected in parallel with associated Time Fuse Links (TFL's) and 1 instantaneous earth fault DAT solenoid Coil.

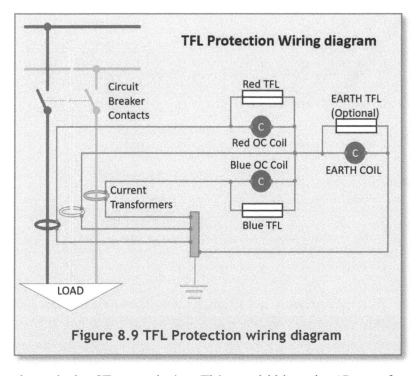

Figure 8.9 TFL Protection wiring diagram

As an example, consider that the circuit breaker supplies an 11,000-volt 1000kVA transformer. Under normal full load conditions approximately 52 amps will flow in the CT primaries. If the CT ratio is set on 50/5 amp ratio, then approximately 5 Amps will flow in the CT secondary winding. As the overcurrent coils are shorted out by the fuses, no current will flow through the coils. Assuming that there was no earth leakage, the red, yellow and blue phases will balance, and no current will flow through the earth fault coil.

In the event of a phase-to-phase fault of, say, 500 amps, 50 amps would flow through the CT secondaries. This would blow the 15-amp fuse links shunting all of the current through the DAT coils, which would operate and trip the 3-phase circuit breaker.

Two overcurrent coils are required to cover all phase fault combinations, i.e. for a R-Y phase fault, the CB will trip on the red phase coil. Under earth fault conditions, there will be an imbalance between the red, yellow and blue phases, and current will flow down the residual path of the CTs and through the earth fault coil. The unit will trip instantaneously. If a TFL is fitted in parallel with the earth fault coil then the tripping time will follow the TFL characteristics. Alternatively, the fuse can be removed.

The time current curves for the primary circuit current can be easily obtained by taking the TFL characteristic rating and multiplying this by the CT ratio. The range of time current curves in Figure 8.10 below show TFL ratings for a 100/5 A current transformer. Do not forget to add the circuit breaker operation time, for further details see section 8.4.9 Relay Error and Switchgear Operating Time.

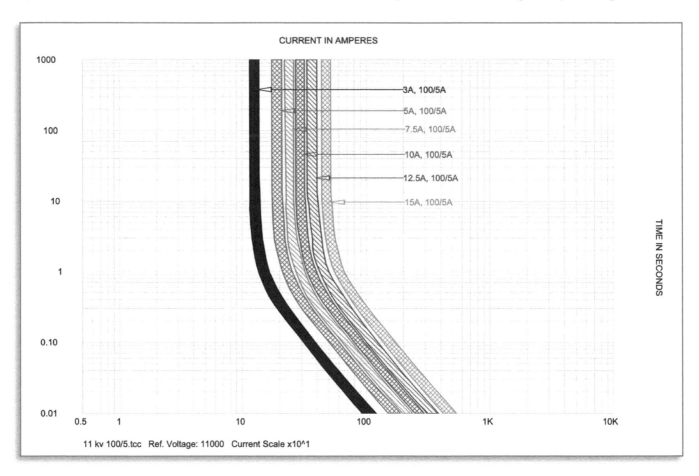

Figure 8.10 TFL Time current curves

8.4.8 DC Tripping Battery Supplies

Locate and check that the tripping battery for the switchgear is working properly and also that it has been maintained. Tell-tale signs of poor maintenance are; no maintenance records, low electrolyte level for non-sealed systems or that battery connections are corroded. The duty holder should be advised that the battery manufacturer's operation and maintenance instructions should be followed, and in particular, the recommended charging rates should be adhered to. I would suggest that the provision of battery monitoring and alarms to warn of imminent failure of a battery system is essential. Absence of this facility ought to be flagged up to the client as a major non-conformance. Remember, that for powered protective relays, no tripping battery = no arc flash protection!

8.4.9 Relay Error and Switchgear Operating Time

Do not forget the relay error and switchgear operating time. Trip times are made up of relay reaction time + trip coil delay time + circuit breaker mechanism operating time. All relays have errors in timing which have to be declared if manufactured and tested in accordance with IEC 60255. This will be in the manufacturer's data sheet, but the following table lists typical values of timing errors for standard IDMT relays.

	Relay Technology			
	Electro-mechanical	Static	Digital	Numerical
Typical % Timing Error	5	5	5	5
Overshoot Time (s)	0.05	0.03	0.02	0.02
Safety Margin (s)	0.1	0.05	0.03	0.03
Typical Grading Margin (s)	0.4	0.35	0.3	0.3

Current transformers also have errors although this is prevalent at lower levels of magnetising exciting current. This is of course, less of an issue at higher currents related to most arcing faults.

We also have to factor in the time for operation of the switchgear. This again can be obtained from manufacturer's data, commissioning, or timing tests but as a guide for the purposes of an arc flash calculation a generic allowance can be built in. For high voltage switchgear under 15kV, typical operating times are between 40ms for vacuum equipment and 75ms for oil filled equipment. I have used a conservative figure of 80ms which is valid based upon experience and providing that there is a high level of confidence in switchgear operating times and therefore no further safety margin is built in. For older oil switchgear, I understand that some utility technical engineers may even increase this to 100ms. Where there are any doubts further investigations including timing tests may be warranted.

You must ask for records of maintenance and make the point that trip timing tests be carried out if there are any doubts about the frequency of maintenance and or trip timing tests.

8.4.10 Low Voltage Circuit Breaker Characteristics

Low voltage circuit breakers will generally be one of five generic types.

a) Miniature circuit breaker – up to 125 amperes single or multipole.
b) Moulded Case Circuit Breaker - up to 1600 amperes.
c) Air Circuit Breaker with separate protection relay – up to 6300 amperes
d) Air Circuit Breaker with integral protection relay – up to 6300 amperes
e) Air Circuit Breaker with integral hydraulic/magnetic protection relays.

8.4.10.1 Miniature circuit breaker

The vast majority of miniature circuit breakers that you will come across are built to IEC EN-60989 and IEC EN 60947 that have thermal magnetic protection mechanisms. The following diagram Figure 8.11 shows the typical pickup and clearance time current curves for the circuit breakers manufactured to these standards. No further allowance is required for breaker clearance.

As you can see, the maximum instantaneous setting is for a type D breaker under IEC 60898 is 20x the current rating (In). For the largest miniature circuit breaker rating of 125 amperes, this will be 125 x 20 = 2500 amperes. This means that all arcing faults under 2500 to 3000 amperes (depending upon the minimum arc current calculation) will clear in approximately one cycle. There may be a possibility that you may come across older types under obsolete standards such as BS 3871 and IEC 898 where the upper limit may be between 10

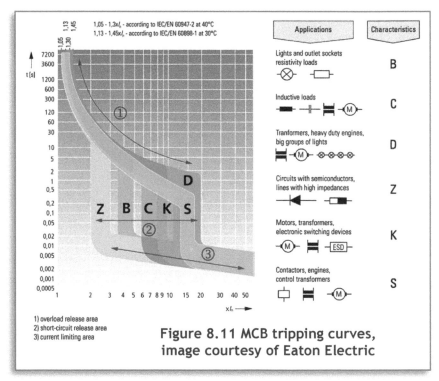

Figure 8.11 MCB tripping curves, image courtesy of Eaton Electric

and 50 times the current rating (In). This is the type 4 miniature circuit breakers under BS 3871 and Type D under IEC 898. This will clearly need checking with the actual circuit breaker type and time current characteristic.

8.4.10.2 Low Voltage Moulded Case Circuit Breaker

Moulded Case Circuit Breakers (MCCBs) are often electromechanical units with thermal magnetic relays producing characteristic curves similar in shape to the miniature circuit breaker. Nowadays MCCBs are supplied with integral electronic trip units as an alternative to thermal magnetic releases, making them much easier to grade with upstream and downstream protective devices. A moulded case circuit breaker is defined as a circuit-breaker having a supporting housing of moulded insulating material forming an integral part of the circuit-breaker. MCCBs can be found up to 1600 amperes. They are very compact and favoured for space saving in factory-built assemblies as outgoing distribution devices although they are sometimes used for incoming circuit breakers from transformers up to 1000kVA. Moulded case circuit breakers up to 630 A can be found with a current-limiting capability. See Chapter 5: Prevention for further explanation.

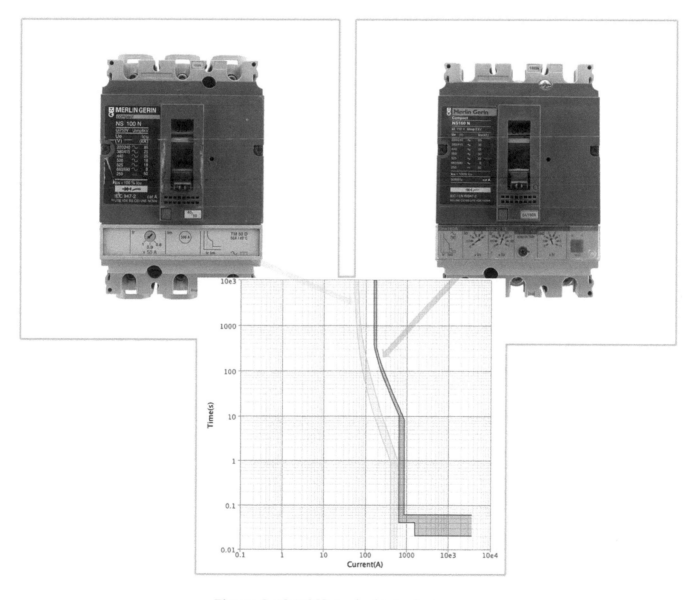

Figure 8.12 MCCB typical tripping curves

The above images in figure 8.12 show similar moulded case circuit breakers, one with a thermal magnetic trip unit and the other with an integral electronic trip unit. As can be seen, the concave slope of the thermal overload section of the thermal magnetic circuit breaker on the left contrasts with the straight lines created by the electronic trip breaker on the right.

8.4.10.3 Low Voltage Air Circuit Breaker

Low voltage air circuit breakers are available up to 6300 amperes and many characteristics are similar. The significant difference between MCCB and ACB is in the rated withstand current with ACBs being of a higher magnitude. Because of this requirement, ACBs have more volume to be able to withstand a high over currents over a greater time. Some older ACBs require maintenance in situ and even modern units require post fault inspection of arc chutes and contacts according to manufacturer's instructions so that is a question to be addressed when collecting data. The vast majority nowadays, tend to have internal electronic trip units built in with multifunctional capability for various protection schemes and monitoring/communication facilities. Carrying out arc flash data collection will throw up other circumstances where external relays such as IDMT protection units and also integral hydraulic magnetic or oil dashpots are present. There are also circumstances where external current transformers have to be taken into account.

8.4.10.4 MCCB and ACB Integral Trip Units

Figure 8.13a is a good example of an integral electronic relay as it shows various overcurrent parameters that are found on MCCBs and ACBs in accordance with IEC 60947-2. It also declares the current transformer ratios which is not usual for integral relays of this class.

The following is a description of the functions of the relay in Figure 8.13a all of which must be collected at the data collection

Figure 8.13a Electronic relay

visit and of course compared to data that may have been received from previous protection studies etc. As you can see for this example, there are frame settings and CT ratios that do not always appear on LV circuit breakers of this class but are important.

- Base current referred to as I_o. This is the same as sensor or circuit breaker rating current usually referred to as I_n on most other units.
- Long-time delay current setting LTD (sometimes referred to as I_1 or I_r and are multiples of I_o or I_n).
- Long-time delay time setting (often referred to as t_r or t_1).
- Short time delay current setting STD (often referred to as I_2).
- Short time delay time setting STD (often referred to as t_{sd} or t_2).
- Short time I^2t setting. This is usually either on or off or sometimes in or out.
- Instantaneous Pick Up current setting INST (often referred to as I_i or I_3 or I_m).
- Instantaneous Pick Up time setting (often referred to as t_3).
- Base current setting (may be a current or a multiple of I_o or I_n).
- CT ratio, not often declared, in this case 1250/5 A.

Base current may be a setting related to the circuit breaker frame size. So, if the frame size of the actual breaker is 1600A (I_n) then the relay current may have a setting of 1, 0.9,0.8 (or 1600A, 1440A or 1280A for instance) In our case the base current is quoted as I_o =1000A. The base current setting of the relay is then 1 or 0.8 or 0.63 times the CT primary rating (I_{ct}).

The following are images of electronic relays which have similar functions to the example given above. I mentioned how important it is to compare protection data collected with previous protection studies and the reason is as follows. In most cases all that is required to alter the protection characteristics is a screwdriver. Although the above relay was behind a hinged door (unlocked) and in a substation, there are many instances where the equipment is open areas such as corridors and does not have a door or a cover. You may also want to explore how future tampering can be avoided for obvious reasons, one of which is that the safety of personnel may be compromised.

To make sense of all the above information, Figure 8.14 shows how the settings impact on a typical over current characteristic for a trip unit on an IEC 60947-2 circuit breaker. I have displayed the basic overcurrent functions and have purposely left out earth or ground fault settings that may be available.

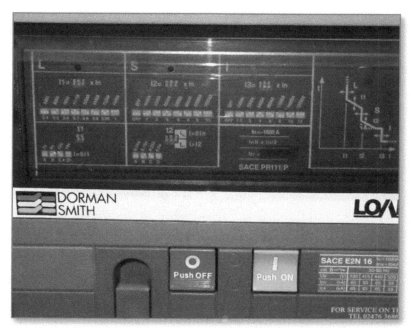

Figure 8.13b Electronic relays

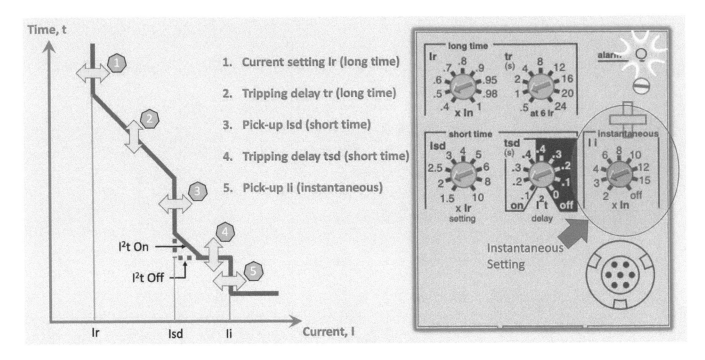

Figure 8.14 Typical settings for integral trip units

The next generation of electronic trips that are available today have much more functionality. They provide more functions of protection but also metering, diagnostics, communication, and remote operation. This may or may not present problems in data collection because gone are the sliders, dip switches and control knobs and these have been replaced with a liquid crystal display and password(s) and/or a key. In Figure 8.15, this illustrates the new layouts.

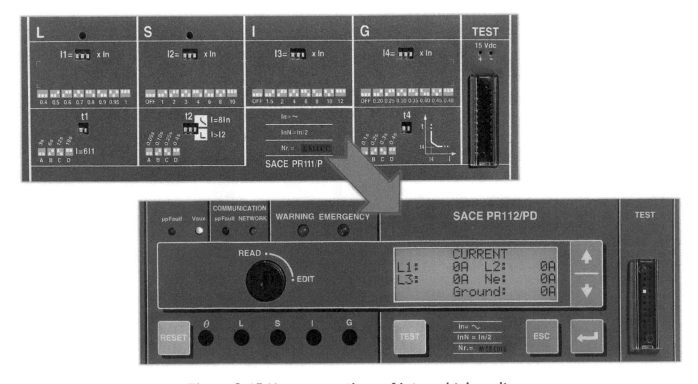

Figure 8.15 New generations of integral trip units

8.4.10.5 Hydraulic Magnetic (Oil Dashpot) Relays

Hydraulic magnetic or oil dashpot over current relays were used extensively on low voltage switch boards and very often as the main incoming circuit breaker from the distribution transformer. As such, they would have a key role in protection of a high value asset and are frequently installed to protect against elevated levels of incident energy for the facility. A diagram of a typical oil dashpot relay is shown in Figure 8.16. As can be seen, a supply from the circuit current transformer is fed onto a current coil which attacks a piston through dashpot oil. Holes in the piston together with the viscosity of the oil acts as a damper and gives the characteristic curve of the relay. The threaded shaft gives further adjustment to the time current characteristic.

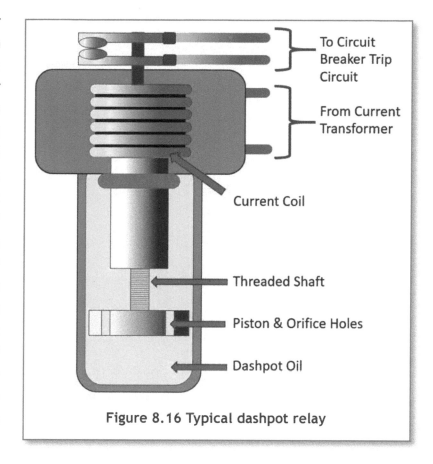

Figure 8.16 Typical dashpot relay

The trouble is, with all my experience in the industry, I have rarely found oil dashpots that have been maintained either properly or recently. The oil in the dashpots have been found to resemble sludge rather than viscous mineral oil in some instances. They are also notoriously difficult to obtain the time current characteristics although I have come across generic curves that have been reproduced by design software companies. From a protection point of view, it is quite likely that you may want to disregard them, unless they have been maintained recently and also the tripping characteristics have been proved by way of current injection. If this is the case, then the commissioning report is often a very useful document in determining the disconnection time for an arcing current. A remedy for the duty holder is to have the circuit breakers maintained by a specialist company or to have a new unit retrofitted.

8.4.10.6 Fuses

In electrical power distribution, where fuses are used they tend to be of two particular types in Europe. They are both high rupturing capacity (HRC) fuses to the German NH DIN fuse or the British Standard BS88 fuse and both fall into the category of current limiting protection devices. See Figure 8.17. The is more information given in Chapter 5: Prevention. The fuses are not physically interchangeable, but the characteristic curves are similar. Both NH DIN and BS88 fuses have been around for many years and have a proven reliability record and have remained popular with specifying engineers to this day because of their high fault current handling and minimal energy let-through (I2t). They tend to be the favoured type of protection for utility distribution but are commonly found in industry particularly for motor protection.

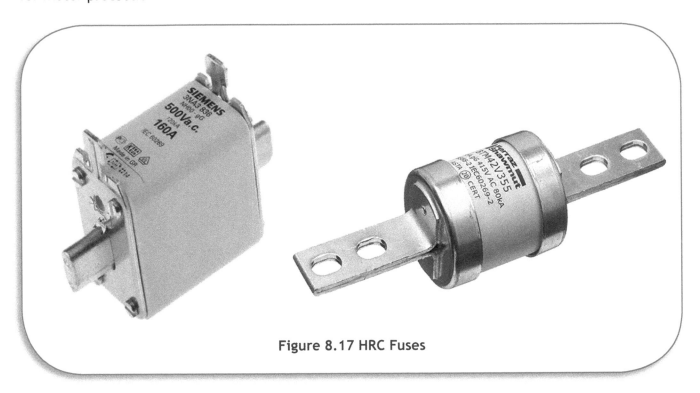

Figure 8.17 HRC Fuses

In terms of data collection, a power interruption may be required in most cases to inspect the fuse body unless separate data is available. Even then, there has to be confidence in the fuse type as motor fuses may be a higher rating in a smaller sized body. There are NH DIN and BS88 fuse interactive time/current characteristic curves appended to this guide for information which include allowances for total clearance times. Other fuse characteristics are generally available by web search and from the main international fuse manufacturers. Be sure to understand what the manufacturer's time current curve characteristics are displaying. Those curves which display pre-arcing times are unlikely to include any allowance for fuse tolerance or total clearance time.

8.5 Low Voltage Cables and Distribution Busbars

This is where the real detective work starts and the tools that are available with this guide will help. If the records are good and you have cable schedules to hand, then that is a real bonus. Even then, if the records are generated by "in service" testing and inspection regimes, then do not expect the information to be accurate for the purposes of arc flash calculations. They will, however, be a good guide. If the records are "as fitted", then this is usually generated by a design engineer and whilst the cable sizes and types are likely to be accurate, the cable lengths may not be. The message is caution, sense checking and inspection.

How long is the cable? The length of the cable is quite important if it is low voltage, but accuracy of a high voltage cable is not as critical to the calculation of incident energy. Having possession of accurate site plans is really beneficial. Cable management systems can be marked up onto the plans and once the accuracy is validated, this can save a great deal of work by scaling off cable runs. Confidence is the key as the final figure is going to be used to calculate the fault level which will determine the incident energy. Where this confidence is in question, then a good idea will be to run 'what if' scenarios where the incident energy is calculated for different lengths. For high voltage cables, an increase in length and therefore circuit will normally result in a decrease in incident energy. For low voltage cables, the opposite may be true due to the fault level paradox explained in Chapter 4: Prediction.

Cables with a plastic over sheath whether PVC or XLPE will usually have the information embossed or indented onto them for identification purposes (See below, Figure 8.18). In addition, there is sometimes a marker on the outside of the cable giving distance in metres which can be really useful. So, you just subtract distance at the receiving end from the distance at the source and you have the cable length. It might be a good idea to have some chalk handy to rub onto the sheath of the cable to make any marking easier to read. White tape or paper and a soft leaded pencil will often do the trick as well.

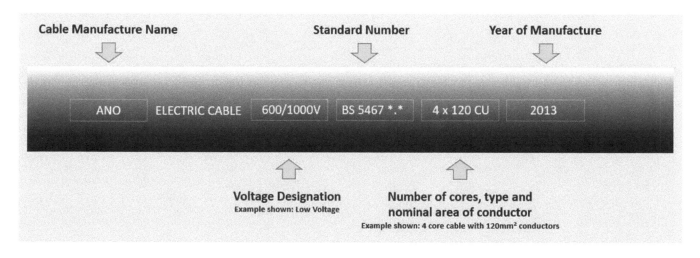

Figure 8.18 Cable markings

The next issue is to determine the type of cable and what material the conductors made from. For modern cables that are embossed on the sheath as shown in figure 8.18 above, that can be determined very easily. I have known times when the whole data collection experience has been made so much easier because of clear cable markings. There have been other situations because of the age of the site where this information has been more difficult to collect. I might just add here that I have come across a large site where there had been counterfeit cables installed which I suspect was more common at some time in the past. Not only no markings, but the XLPE insulation was wafer thin in places so please be mindful that if a cable looks insulated in a cable box, it might not be.

The cable diameter is a good pointer towards the size of the cable if you know what the conductor material is, and manufacturer's data will help determine conductor size from the outside diameter of the cable. Also beware, that I have come across reduced neutral conductors as part of a four-core cable but, on enquiry to local cable manufacturers, there does not seem to be a large market nowadays. Another strategy is to examine the terminations as a last resort, dead of course. The terminal lugs will very often have the core size stamped on the lug. As always, sense check information like this against other factors just in case oversized lugs may have been used.

The way in which the cable is installed, also has a significant effect upon the impedance of larger cables. It should also be pointed out that the larger the cable, the more that the reactance will contribute towards total impedance. For instance, the X/R ratio tends to approach one for copper cables at about 300mm^2. Above that cable size, reactance has a greater contribution to overall impedance. Single core cables in particular can have large variations in reactance dependent upon installation method. For instance, single core cables laid flat and spaced can have reactance values that are double those in trefoil formation. That is very significant for transformer tails as the X/R ratio can be as high as four times. A similar effect of increased reactance can be experienced for cables installed in trunking. As can be seen, the reactance values can vary widely with different installation methods and any electrical engineer who is carrying out retrospective system studies needs to be aware of this fact.

Cable Installation Method	Composite 3 phase Cable	Trefoil or Delta Single Core Cables	Single Core Cables Touching	Single Core Cables Trunking	Single Core Cables Spaced 2X or 4X Radius
Diagram	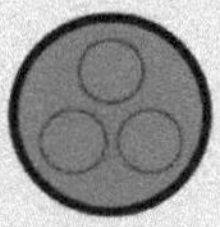			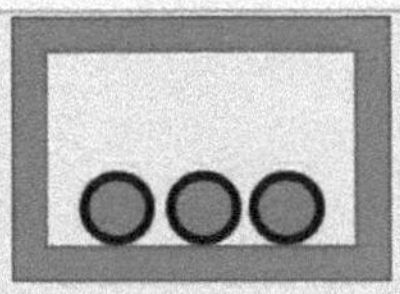	
Generic Average Reactance (milliohms/metre)	0.08	0.085	0.095	0.135	0.145 2X Radius 0.190 4X Radius

Figure 8.19 Average reactance values - source Schneider Electric

Data from distribution busbar systems can usually be found from manufacturer's data sheets. Rarely can the impedance data be determined from busbar nameplates and the extent of useful information will be the current rating and voltage. Even the busbar material and cross-sectional area will be missing so applying generic data will be difficult without reference to the manufacturer's data sheets. The good news is, that most present-day suppliers provide comprehensive data on their products and information such as resistance, reactance, current carrying capacity, short circuit withstand and voltage drop.

8.6 Transformers

Transformers usually have data stamped on the nameplate and you will be unlucky if the minimum information to calculate symmetrical fault level at the secondary terminals is not available. See Figure 8.20 as an example of a typical transformer nameplate.

By minimum information for distribution type transformers, I mean the transformer type, cooling (forced/natural), insulant (oil, air or synthetic fluid), size (rated kVA), primary and secondary voltage and most importantly the voltage impedance as a percentage. Larger transformers may need further information such as automatic tap change mechanisms. As part of the visual checks, I think that it is important to report on the physical condition of the equipment with an eye open for cable box compound leaks, insulant fluid leaks, unusual sounds or smells, security of cable box covers and security locks on off load tap change mechanisms. All these issues could have an effect in the likelihood of a future arcing fault. The transformer X/R Ratio is unlikely to be stamped on distribution transformer nameplates but if you manage to locate the manufacturer's test sheet then that information will be available. Failing that the prospective short circuit current calculator tool available with this guide (ea-guide.com) has generic X/R Ratios for distribution transformers up to 3000 kVA.

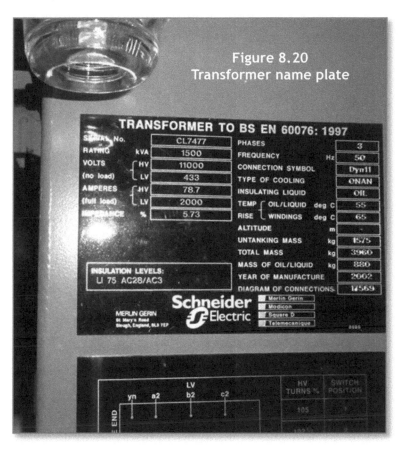

Figure 8.20
Transformer name plate

8.7 Motors

We know that induction motors can contribute to fault level and need to be taken into account when undertaking short circuit studies. When undertaking complex short circuit studies in accordance with IEC/EN 60909-0 standard, the effect of motors needs to be taken into account if the motor outputs exceed 25% of the total installed. The information that will be required as a minimum will be motor type (synchronous, induction), power factor, efficiency, motor starting, speed control, locked rotor current, sub transient reactance value and X/R Ratio.

Incident energy is affected by motor contribution because the increase in fault level will cause more arcing current to flow for the duration of the contribution. The following simplified diagram Figure

8.21 shows how the contribution from the motor can be approximated by taking the incident energy from the motor and adding it directly to the incident energy from the utility up to the point that the circuit breaker (CB) opens and disconnects the circuit.

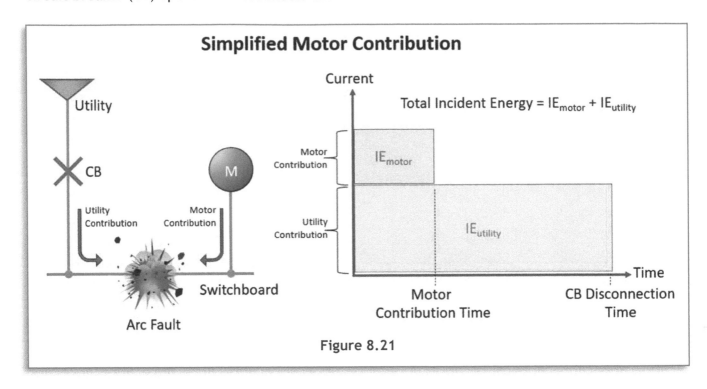

Figure 8.21

The motor incident energy calculator which is available with this guide (ea-guide.com) uses this approximation to allow the user to arrive at a conservative estimate of motor contribution. In terms of data collection, you will need the motor size (kVA, kW or full load amperes). By entering a motor contribution time, which is normally within 100 milliseconds, and an estimate of magnitude of current, often quoted as 4 to 6 times full load current, this will allow for the total incident energy to be estimated.

There was a suggestion in the 2002 edition of the IEEE 1584 Guide for Performing Arc Flash Hazard Calculations, that motors above 50HP (37kW) could be ignored as part of an arc flash study. No such guidance appears in the latest edition and this is where there may be some engineering judgement. You may wish to add all connected motors regardless of size and a acceptable approach will be to add all the motive power and apply it is one large motor. In addition, electronic variable speed drives (VSDs) may be ignored as they block fault contributions from a motor except where the VSD is in bypass mode or where the origin of the fault is in the VSD panel.

If the motor contribution estimate is to determine arc flash incident energy only, then estimates based upon doing "what if scenarios" is, in my view, justified. However, if the objective is to carry out an arc flash study in compliance with the standard IEC 60909-0:2016 (Short-circuit currents in three-phase a.c. systems - Part 0: Calculation of currents) then detailed guidance is given about which motors can be ignored.

8.8 Working Distance

There is good reason when carrying out a hazard study for a facility to try to standardise on the working distance. This is normally expressed a working distance at low voltage and a second working distance at high voltage. It can reduce confusion when comparing the incident energy at one piece of electrical equipment with another. For instance, you may say that panel A is 10 cal/cm² at a working distance of 500mm and panel B is 8 cal/cm² at a distance of 600mm. The fact is that panel B is likely to have a more severe incident energy if the working distance is the same. But actual working distance can change, as we know from this guide small reductions in working distance produce large increases in incident energy. The calculators available with this guide (ea-guide.com) allow for dynamic risk assessments to be undertaken for variations in working distance where different tasks create changes to how close a person is to the arc.

In Chapter 11: Electrical Utilities, there are examples of this, and the image in Figure 8.22 shows one such case. In the IEEE 1584 Guide for Performing Arc Flash Hazard Calculations, there is a standard default working distance of 914mm for operations on 15kV switchgear.

As can be seen in Figure 8.22, the operator is manually racking in a vertical isolatable 11kV oil circuit breaker in a training workshop, hence why parts are cut away. Parts of his body can be seen to be approximately 200mm away from the busbar spouts. This example graphically illustrates that working distance should not be a one size fits all and anyone carrying out a study should be able to recognise these variances and report on them.

Figure 8.22 RACKING AN OCB

Figure 8.22

8.9 Modes of Operation

It is preferable to get the modes of operations established before leaving site as this could have an effect on the final study. For instance, I remember one site where the client had a maintenance regime which involved the paralleling of transformers. This was achieved by overriding interlock systems that should have prevented such an operation. As it happened, the duty holder did not see the issue as an important one as the duration of parallels were for short periods. By modelling this scenario, not only did it reveal that the low voltage system was overstressed but incident energy levels became extremely high. The other factor to consider was that the likelihood of faults tended to increase post maintenance. This all came to light by discussion on site and by examination of the equipment concerned.

The main thing to remember is that things change, and these changes need to be modelled as scenarios into the final study. It is also good to sense check any assumptions that you may have made with the duty holder so that it becomes a consensus of opinion. The typical scenarios might be as follows.

- Scenario 1 - Base Model – Utility supplies intact.
- Scenario 2 - Utility supplies in maintenance mode.
- Scenario 3 – Emergency Standby generators running in island mode.
- Scenario 4 – Emergency Standby or CHP generators running in parallel.
- Scenario 5 – Transformer parallels and so on.

8.10 Consolidation of Data

When considering large studies, it is prudent at this stage to take stock and check off all the data collected. There may be a list of issues that are unresolved, and if they remain that way, will threaten the smooth expedition of the final study. There will be a great deal of data in a myriad of different formats, so it is important to get it organised. The following is a suggested structure to use when consolidating the data.

1. Circuit and equipment data sheets. The data sheets need to be formalised and checked. Missing data can be spotted easily if following the guidance given previously in this chapter.
2. Photographic evidence. This will need cataloguing perhaps by substation and may need to be checked by the client for commercial sensitivity.
3. Assumptions. Some assumptions may have come about by personal engineering judgement or by agreement with the client/duty holder. It is essential that this is recorded and formally agreed. Examples of assumptions are; modes of operation, working distance, competence of workers, equipment maintenance or inspections and equipment condition (see previous section). This ought to include the actions required to address future changes to network configuration.
4. Safety Related Information
   4.1 Hazard report forms that have been issued to report serious hazards that are spotted to the duty holder.
   4.2 Design and Installation hazards that have been experienced during the data collection process.
5. Register of all information received from the site either formally or informally.
6. Action planner. Outstanding issues or requests for information will need to be captured in an action planner. This will drive the progress of the project.

When the consolidation process is completed, it is essential that the duty holder/client verifies the data. It is suggested that a package of information containing all of the above plus initial single line diagrams is transmitted to sign off the data collection stage.

Learning Points

> It is vital that the planning and preparation phase is undertaken with some rigor if the overall data collection process is to be successful.

> Get onto the electricity supplier or third-party supplier in respect of the characteristics of supply at the earliest opportunity for a successful outcome.

> Use the analytical tools and calculators with this guide throughout the data collection process to determine worst case scenarios when doubt exists about data interpretation.

Chapter 9

Electrical Duty Holders

The scope of this chapter is help duty holders to discharge their responsibilities specifically around the management of the arc flash hazard. Electrical duty holders are classed as anybody who works with electricity and their duties relate to all those matters that are within their control. For the purposes of this chapter, the electrical duty holder is the one who has prime responsibility and the resources to provide a safety management system for any facility. So, if you identify yourself as that person, the following are specific measures that you can take to help with your decision making in respect of the arcing hazard.

In EN 50110-1:2013 Operation of Electrical Installations, the duty holder will be termed as the "person responsible for an electrical installation". This person is defined as the nominated person with the overall responsibility to ensure the safe operation of the electrical installation by setting rules and organisation or framework. In Chapter 6: Process, Policies and Procedures, the need is identified for establishing the responsibilities of stakeholders for implementation, planning, control, authorisations, audit and review. This should identify duty holders in respect of an electrical safety management system encompassing safe procedures, safe people and safe places as shown in Figure 9.1.

Safe People
Competence, information, training & instruction. Supervision & control.

Safe Procedures
Risk assessment approach, specific rules with clear responsibilities & authorisations.

Safe Places
Well-designed & documented installations, maintenance and confidence in protection systems.

Figure 9-1
Building Blocks of Electrical Safety Management – Theme courtesy of Guardian Electric

Figure 9.1

9.1 Safe Procedures, People and Places

It is against the above model that we can frame some of the key actions that we need to take as duty holders to manage the arc flash risk. Whilst not exhaustive, (depending upon the organisation and environment), these key actions are detailed in the following arc flash specific checklists. Note that the safe procedures, people and places model overlaps indicating that these facets are mutually dependant. To give an example, a safe place of work depends not just on the original safe design of the electrical system, it is also dependent upon having competent persons maintain its condition following suitable procedures to keep it that way.

9.1.1 Safe Procedures Checklist

a) **Risk Assessment.** This takes into account the 4P approach. In other words, to predict the level of harm from a possible arc flash and then to prevent from it causing harm. More specifically, we will identify the hazard and those at potential risk of harm, estimate or predict the severity, eliminate or prevent the risk and finally by mitigation protecting those at risk of harm.

b) **Approval of Electrical Test Equipment, PPE and Tools.** Formal control and approvals process for all electrical test equipment, insulated tools, locks, labels, insulating tools, matting, shrouding and electrical PPE.

c) **Clear responsibilities and authorisations.** Individuals for any high-risk activities should be identified with a clear list of responsibilities and then authorised in writing.

d) **Establish dead working as a principal requirement.** Work on or near live exposed conductors should rarely be permitted. Plan and programme the work to allow work to be carried out where possible with the equipment dead.

e) **Specific controlled circumstances for live proximity work.** Be clear about which activities could be sanctioned and under which circumstances. Employ further control measures to meet the risk.

f) **Clear instructions on where live working is strictly forbidden.** Be very clear about which live working tasks are forbidden.

g) **Monitoring and Review.** Where reliance is put on procedures and protection in the management of the risk, there is a high need to ensure that those controls are working properly and reviewed accordingly.

9.1.2 Safe People Checklist

a) **Information, training and instruction.** A good risk assessment methodology will assist greatly in providing awareness, information and an understanding of the hazards.

b) **Competence assessment.** Personnel have a key role in arc flash risk management. All the factors that are outlined in Chapter 6: Process, Policies and Procedures about competence assessment, will help to establish safe people and prevention strategies. In addition, to have named electrical safety ambassadors at the level of operative works well to achieve focus and to influence colleagues and contractors.

c) **Contractor competence.** Make sure that all contractors are competent at tender evaluation. From my experience, there are a disproportionately high number of arc flash incidents among contractors. As highlighted previously, control of contractors specifically down to the level of confirming the competence of their personnel is required.

d) **Specific authorisations.** As mentioned previously, operatives who have clear boundaries, and are authorised through a rigorous assessment of competence, are less likely to carry out high risk activities involving live equipment.

e) **PPE.** The provision of personal protective equipment is a last resort but it is necessary that correct PPE is provided and worn. See Chapter 7: Protection. Any PPE must not create other hazards, such as severe sensory deprivation so this is why the 4P approach is so important.

f) **Supervision and control.** Wherever mitigation procedures are put in place, whether PPE or rules, it is essential to ensure that these measures are being practiced by personnel. Supervision and leadership backed by auditing processes will ensure that what is put in place in theory, actually happens in practice.

9.1.3 Safe Places Checklist

a) **Well-designed installations.** Having installations that are designed in accordance with best practice and take into account the techniques highlighted in Chapter 5: Prevention.

b) **Well maintained installations.** An electrical maintenance program is required to keep electrical equipment in good working condition so that it can reliably and safely operated within its design criteria. Improper maintenance of equipment can increase the likelihood and/or contribute to the severity of an arc flash.

c) **Confidence in protection systems.** Testing and maintenance of circuit breakers, protective relaying and associated equipment on a regular basis. When a protective device such as a circuit breaker or relay is not properly maintained, the likelihood of it operating slower (or not at all) increases, which would also increase the duration of exposure to the arc flash.

d) **Organisation of Documentation.** Having an organised library incorporating design information, test data, commissioning information, grading and fault level studies, technical manuals, asset directory, manufacturer's operating manuals and instructions, periodic test and inspection reports is essential. This is often overlooked as a key ingredient in electrical safety management.

e) **Fires from electrical arcing.** Although not the main subject of this guide, there is evidence that the largest cause of fires in non-residential premises is due to electrical faults and an estimated 50% of these are arcing faults. For high-risk systems or areas consider arc fault detection as enhanced protection and most importantly, well designed and maintained electrical systems.

f) **Carry out a system study including an arc flash hazard assessment.** This can be the catalyst for addressing many of the points detailed above.

9.2 System Study

The last item on the above checklists is about carrying out a system study including an arc flash hazard assessment. This section gives advice on how to recognise whether the information that you have on your electrical system is sufficient and if not, how to improve or correct the situation.

In my view, a study should not be with the single aim of creating incident energy levels in order to specify PPE. To carry out a study of an electrical system provides a perfect opportunity to take stock and to establish whether or not the design of the electrical system matches the present condition, and also to highlight serious deficiencies in operation and maintenance. One such example of what is meant here, is the highlighting of unauthorised modifications to switchgear and electrical panels which, effectively convert a safe piece of equipment into something that is unsafe and potentially lethal. I remember creating a gallery of horrors from photographic images, where such modifications were widespread on one particular site. The embarrassment of the duty holder was palpable as we went through slides showing numerous unauthorised modifications to equipment. One was almost comical, showing a five-gallon chemical drum that had been cut in half to replace missing insulation over an internal process panel isolator.

9.2.1 Expectations from a system study?

A network diagram or a single line diagram. An accurate schematic of the electrical system is essential for a high voltage system but is also important for high power low voltage networks and is a reference document for the planning, coordination and control of work activities.

Fault level study. A fault level study will give prospective fault current level and information about the capability of switchgear. Things change over time meaning that fault currents can change because of characteristics of the electricity supply, embedded generation and system configuration. Whilst we are looking for the worst-case fault current available at the various connected electrical equipment, we need to also ensure that we understand the normal fault currents in order to determine incident energy levels. The worst-case fault current will determine the strength and capability of connected equipment to ensure that it will not fail during a fault clearance. The study will take into account mains connections, embedded generators and also other fault contributors such as induction motors.

Protection Coordination Study. Any changes to characteristics of the electricity supply, embedded generation and system configuration can require that protection arrangements are reviewed. As the name suggests, the study will ensure that the electrical system will react to faults in a predictable manner and will determine the length of time that a fault takes to clear. In order to calculate incident energy, the coordination study is necessary to determine the clearance time of arcing currents.

Arc Flash Incident Energy Study. The arc flash study is designed to calculate the arc flash hazards at a distance called the arc flash boundary and the incident energy that a worker could be exposed to when working on or near live equipment. I have calculated that the amount of effort that is required to carry out an arc flash study is about 10% of the total when conducting the creation of a single line diagram, fault level study and a protection coordination study. That is, bearing in mind that any self-respecting engineer would advocate the latter three outputs to be essential to electrical system management.

9.3 The Risk Gap

The reason why I have segmented the process in this way is to give flexibility to the duty holder in how they can gain compliance with best practice. Going back to basics, we have a known hazard, called arc flash, which should get attention in the same way as any other potentially serious hazard. What I would hate to think, is that a duty holder would fail to carry out a risk assessment purely because of a mistaken belief that the only way to proceed is by commissioning a full arc flash study. That could be a costly way to kick the can down the road and to miss the overall point about risk assessment.

In the UK, the HSE expect *"that duty holders, in turn, will adopt a sensible and proportionate approach to managing health and safety, focussing on significant risks i.e. those with the potential to cause real harm. In applying the principle of proportionality means that inspectors should take particular account of how far duty holders have fallen short of what the law requires, and the extent of the risks created"*. During visits by inspectors, a risk gap analysis is made to determine the actual risks and measure this against the risk accepted by the law or guidance. The sooner that risk assessments are commenced on a sensible footing, and focussing on significant risk, then the more likelihood of bridging that gap between what is happening now and what should be happening.

Figure 9.2 Risk Gap reproduced courtesy of Guardian Electric

9.4 Choices of Approach to Predicting the Hazard

9.4.1 Option 1. Do Nothing

Not to be recommended although, limited action as a result of a risk assessment is legitimate.

9.4.2 Option 2. Use the Guide

There are several advantages to using the guide which are:

- The guide will identify significant risks quickly.
- It will provide scalability, flexibility and be a steppingstone to a more complex approach, should it be necessary.
- It is a very low cost and accessible alternative and inclusive for more members of the engineering team.
- Can easily be adopted into dynamic risk assessments.
- Provides a source of training and information.
- Save Money by:
  - Low initial cost of model.
  - Optimising PPE.
  - Reducing training costs.

- ○ Cost reduction Strategy and commercial contracts and vendor advice should a more complex approach be required.
- ○ Identify and group low level circuits into categories that can be excluded from complex arc flash studies.

9.4.3 Option 3. Outsource a Study

If you have concluded that the correct course of action will be to outsource a system study, I hope that it is based upon some of the advice within the guide. Once this decision has been made, the following suggestions may help in obtaining the best possible outcome when commissioning the system study.

Type of software. There is further guidance on complex modelling software given in Chapter 13: Complex Software Guide. Something that the duty holder may need to watch out for is where complex software packages are being used by the company elsewhere in the world. So, if there are other sites within the group, maybe they have gone through a similar learning curve that you may not have to repeat for the sake of a phone call. There may also be a commercial contract in place that may lead to some significant discount. Another issue is the compatibility of the software packages with software already in use perhaps for importing or exporting data.

Which circuits are going to be included? This is where some huge savings can be made. In terms of effort, a 32-ampere low voltage final circuit takes nearly as much time to survey as a 3200-ampere transformer air circuit breaker. Circuits below 100 amperes will probably be far more abundant and outnumber larger power circuits by a factor of 10 plus. By using the guide, you can determine a cut off limit for low voltage circuits which will drastically reduce the number of circuits in the overall study. For instance, if you were look at the tier 3 calculator for devices less than 125A calculator the incident energy at a working distance of 450mm, devices below 100 amps will clear circuits very quickly except for very low fault levels. These low incident energy level circuits can be confirmed from the calculator, then grouped and then categorised as being below 1.2cal/cm². As such they can then be removed from the overall modelling study which will reduce overall costs.

Remember that a burn will still be possible from an arc flash incident as the hands may be closer to the arc when testing for instance. This is where I would always advise on the use of insulating gloves with leather over protectors regardless of circuit size for testing. Use of the tools with the guide will allow you to create a cut off criteria in order to identify which circuits will be excluded from the study.

Choosing a Competent Contractor. There are vendors that are available who claim to be proficient in arc flash studies and it may help to ask a number of questions to ascertain the correct credentials. This is in addition to competence assessment that is covered in Chapter 6: Process:

- Experience
  - ○ What arc flash studies has the prospective vendor completed?
  - ○ Are there suitable references from a previous client?
  - ○ Have they had experience on similar sites to your own?

- European approach

    - Can they demonstrate a European approach to electrical safety?
    - Ask to see outputs from studies, do labels follow European signs and symbology rules?
- Electrical safety credentials
    - Are they providing the study from a purely academic standpoint?
    - Do they provide electrical safety advice alongside the numerical output from the study?
- Training
    - Can they provide and include electrical safety training which incorporates arc flash to your employees and contractors?
- Data Collection
    - Has the vendor a process for data collection?
    - How does it compare with the planning aspects of Chapter 8: Data Collection?
    - What safety precautions are planned for inspections near live equipment?
- Software - There is further guidance on complex modelling software given in Chapter 13: Complex Software Guide
    - Is their software package being used by the company elsewhere in the world?
    - Do they provide a free viewer or is the output restricted to written reports and spreadsheets?
    - Does the system model and data belong to you or the vendor?
    - What is the cost of purchasing a copy of the software for future circuit additions?

9.4.4 Option 4. Buy Software and DIY

I know individual duty holders who have purchased their own software and have successfully carried out studies on their electrical networks. There is, however, a learning curve if you have not had previous experience with electrical modelling software. If this is a course of action that you are contemplating, the following questions may help in the decision:

- Have you read Chapter 13: Complex Software Guide?
- Have you tried the software? You can test drive most software packages for a month. It is worth it as the user experience is very different between different companies.
- Do you know the true cost of the software? I have explained how this can be mitigated in my Chapter 13: Complex Software Guide. You need to factor in the ongoing support.
- Have you looked at leasing the software?
- How many licenses do you need? If the software is going to backup dynamic risk assessments, then additional licenses will be required.
- Training. Do you know what the training costs will be? If you are a newcomer to complex system studies, it is important to get training which can be quite expensive. I would recommend that this cost is factored in before making a purchase.
- Is local support available? This is quite important for the newcomer in particular when applying software to local switchgear norms and local regulations.

9.4.5 Option 5 Long Hand Calculations

All the calculations that are used in this guide or from complex modelling software are derived from the IEEE 1584 Guide for Performing Arc Flash Hazard Calculations or the ISSA Guideline for the Selection of Personal Protective Clothing when Exposed to the Thermal Effects of an Electric Arc. Practical training is available to help with the understanding of the IEEE 1584 calculations, and much more, from Jim Phillips of brainfiller.com. Jim provides students with worksheets that will allow them to calculate incident energy and arc flash boundaries from first principles. It is also a great way to learn more about the hazard and can be highly recommended.

9.5 Prioritisation

As I have said may times, arc flash is a hazard that should be treated in the same way as any other. The goal must be to prevent the hazard from causing harm and should there be any residual risk, then to provide protection for workers through lightweight solutions. However, there is no need to wait for system studies to be carried out to prevent injury. My advice is, do not delay, start managing the risk right away or **A.S.A.P.**

9.5.1 The A.S.A.P Principle for Duty Holders

By the ASAP Principle (As Soon As Possible) what I am trying to emphasise is that you do not have to wait for consultants or complex software before you start to reduce risk from the arc flash hazard. You can start straight away. The following ASAP steps will help you to get started.

Awareness: Read the guide, discover the conditions and actions that commonly lead to arc flash and the injury and damage. For operatives this can lead to the step change in attitude that will be required to working safely.

Stop: Stop or avoid those activities that are high risk that arise out of the awareness step. Take decisive action, do not tolerate or turn a blind eye to custom and practice habits that often lead to unsafe working. This step can be implemented very clearly and effectively by issuing simple instructions.

Audit: As I have quoted before; "If you can't measure it, you can't manage it." From experience, equipment and control panel audits can be a very effective way of taking stock and will paint a picture of the effectiveness of existing risk control, housekeeping, maintenance, dilapidation and identification of unsafe practices or equipment. It will also provide a means of determining the arc hazard level very quickly if using the tools provided with this guide. See Chapter 14: Hazard and Severity Calculators.

Plan: Create a plan which looks at a review of policy, procedures, people and places. From the information gathered, look to providing the leadership and strategy to amend rules, asset improvement or replacement and at this stage, decisions can be made to commission an arc flash study if necessary.

To summarise the above:

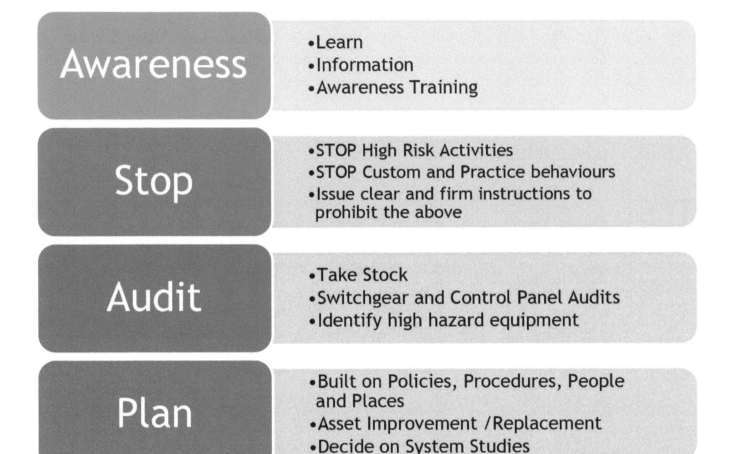

The first two actions, awareness and stop will in my view be the biggest step change that one can make and can often create good returns fairly quickly. The audit stage can provide the catalyst for change by shining a light on existing work practices and equipment and if carried out effectively, will feed into the fourth stage of planning.

The following is an example of a control panel audit form which can be used to as one part of the audit step. As discussed previously, dealing with arc flash in isolation will miss an opportunity to very effectively take stock of equipment which is likely to have the greatest number of interactions and therefore risk in terms of likelihood of electrical incidents.

Example Control Panel Audit		YES	NO
Panel Reference [blank] **Location** [blank]			
Working environment			
1	Is the access/space adequate to allow the worker to pull back without hazard?		
2	Can access doors be opened safely with no obstructions to access safely?		
3	Are the lighting levels adequate?		
4	Is the immediate area free from dust, water or other substances?		
5	Check there NO risk of ignition hazards?		
Equipment and Shock Hazard			
6	Are warning labels and panel references in place?		
7	Can the panel be secured against unauthorised entry?		
8	Is the isolator functioning correctly?		
9	Can the isolator be securely locked in the off position?		
10	Are external connecting cables in good condition with glands and shrouds secured?		
11	Are earth connections for outgoing armoured cables sound?		
12	Are earth bonds on extraneous metal work in place?		
13	Is containment such as conduits, basket and trunking lids free from damage?		
15	Are components including cables free of any signs of damage?		
16	Is the panel free of any exposed cores, corrosion or signs of burning or discoloration?		
17	Is the panel free from signs of water/dust ingress		
18	Are the door seals intact?		
19	Are ventilation fans clean, operational and fitted with a serviceable filter where appropriate?		
20	Is the condition of equipment cables, lamps, bulbs, conduit glands and local switches in good order?		
21	Is the panel IP2X Finger safe?		
22	Is the panel free of holes or unused cable entries?		
23	Is the house keeping inside the panel in good order?		
Design and Utilisation			
24	Has the panel been properly installed and maintained?		
25	Has the panel been used in accordance with manufacturer's instructions?		
26	There are no modifications to the original equipment specification		
27	Are equipment doors and covers closed and secured?		
28	Is it verified that there is no evidence of impending failure?		
Arc Flash			
29	What is the prospective fault level at the panel?		
30	What are the upstream device settings? [blank]		
31	What is the incident energy at the panel @ 450mm working distance?		

Note. This is for guidance only and not meant as a definitive or exhaustive panel risk assessment. This is a visual inspection and not a test and inspection process under wiring regulations. Modifications may need to be made to ensure that it is site specific.

9.6 Hazard Calculators

My approach throughout this chapter has been to start the arc flash risk assessment process without delay. The tools that I have provided will give the means that any competent electrical power engineer to determine the incident energy level at any piece of equipment. This is assuming that the electrical system at your facility is a traditional power distribution system comprises a high voltage point of connection or perhaps a low voltage bulk supply from a utility. In Chapter 10: Service Providers and Contractors, the section on Prediction - Incident Energy, describes a tiered approach to determining incident energy as shown in Figure 9.3 below.

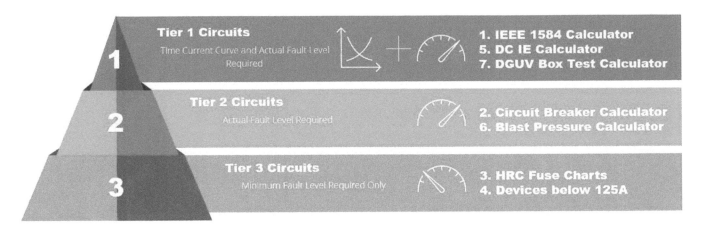

FIGURE 9.3 TIERED CALCULATOR APPROACH

In a generic tiered approach, circuits and associated calculators and tools are grouped as tier 1, 2 and 3. In the pyramid diagram, tier 3 circuits which are the most plentiful in most industrial and commercial facilities. As you can see, the only information that will be required is the type and size of the upstream protection and the minimum prospective fault current level. Tier 2 circuits need actual fault level only and finally, at the top of the pyramid, the actual fault level and a time current curve will be required to return an accurate prediction of arcing current and incident energy. There is also a prospective short circuit current calculator and additional tools to plot time current curves for IDMT relays and European style fuses. Chapter 14: Hazard and Severity Calculations explains all the tools in detail.

Learning Points

➢ Follow the checklists for safe procedures, safe people and safe places.
➢ Identify the risk gap.
➢ Start managing the risk straight away – Use the A.S.A.P. principle.

Chapter 10

Service Providers and Contractors

Having been a service provider, a contractor and a consultant, I feel that I have a good understanding of the difficulties faced by these professionals when entering a site or facility where the electrical safety management system may not be ideal. One of the first hurdles could be doubts over the competence of key appointments such as site electrical duty holder. The site or facility may have suffered from neglect and a lack of maintenance and in fact, your role may well be to improve the electrical installation in these circumstances. This chapter is therefore dedicated to you as a service provider who may be involved in various on-site activities that could be affected by the arc flash hazard. Whether you are an electrical installation contractor, high voltage maintenance company, lift technician, data collector, maintenance technician, refrigeration servicing engineer or a periodic test and inspect engineer, the following will apply whether as guidance, checklist or as awareness of the hazard.

First of all, it is worth reiterating just how important the management of the arc flash risk is to a specialist electrical service provider. I can point to my chapter on myths and mistakes and the times I have seen electrical service providers involved in arc flash accidents and how they could have been avoided. It is surprising just how many common factors there are, and also how many misconceptions of what is reasonable and what definitely is not. The other common thread is the miscalculation of the damage that can be done to a service provider as a result of electrical accidents. There is of course the injury to the person, but there is also the threat of prosecution through the HSE for non-fatal accidents. Then there is brand damage which can be acute for service providers where they rely on larger corporate organisations for their business. Avoiding electrical accidents is a core competence for an electrical service provider and as I point out in my Chapter 12: Myths and Mistakes, a prosecution will stay on record and be visible to your clients. You will also have to declare the fact on prequalification and tender submissions for future work. Not good for business!

By being proactive and knowledgeable about the hazard, this will reinforce your credentials as a competent service provider in the eyes of the client. I recognise that it is not going to be possible for a service provider to be able to turn out an arc flash study on every piece of equipment that is worked upon. You can, however, carry out a suitable and sufficient risk assessment taking into account an estimation of the incident energy and the likelihood of an incident by adopting a practical approach to keep your operatives and others as safe as can be reasonably expected.

10.1 Strategy

When you open that control panel door, you need to know whether it is an opening into the equivalent of a tiger's cage as explained in Chapter 4: Prediction. If it is, then maybe it is better to leave the door closed and well secured. But how will you know and what should you do?

As a reminder, we have a legal duty to carry out risk assessments in the workplace. We are guided that the risk assessments have to include the following steps.

1. The identification of hazards.
2. The identification of those at potential risk from those hazards.
3. An estimation of the risk involved.
4. Considering if the risk can be eliminated; and if not go to step 5.
5. Making a judgement on whether further measures to prevent or reduce the risk needs to be introduced.

Steps 1, 2 and 3 are where we found ourselves in respect of the tiger's cage that we spoke of earlier. Hopefully, step 1 the identification of hazards, will not require the opening of the cage door. This is where a strategy is required which will be written into safety rules and supporting procedures that were covered in Chapter 6: Process, Policies and Procedures. The strategy will make sure that your workers will know what to do from a system of awareness, information, training and competence assessment well in advance of stepping into a potentially hazardous situation. So, where do we start?

10.2 The 4P Approach for Service Providers

In Chapter 3: Risk Assessment and the Four P Guide, I explained the 4P approach which can be applied to service providers and contractors in respect of the arc flash risk.

To recap, the 4P model starts with a prediction of the incident energy and then a prevention strategy follows to stop the arc flash hazard from causing harm. Once the prevention strategy is exhausted, the process step is necessary to provide or update rules, procedures, competent people, information and training. Finally, where there is residual risk, protection has to be provided which may include FR PPE. By now you should be fully aware of the 4P approach and the chapters that deal with **Predict**, **Prevent**, **Process** and **Protect** are detailed in this regard. The following is how the 4Ps are applied to service providers and contractors in addition to the other chapters in this guide.

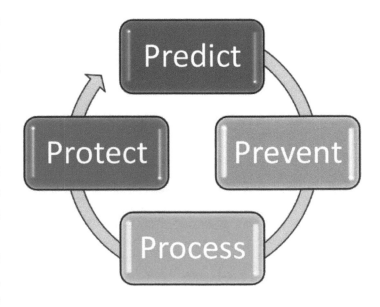

10.2.1 Prediction - Incident Energy

In general terms, the majority of electrical circuits within an industrial or commercial facility, will not have predicted arc incident energy levels that are of an order that could cause serious burns. By serious I mean, when an individual is at a working distance of 450mm or more from a low voltage equipment, a predicted arc flash could cause third degree or full thickness burns. This does not mean to say that no injury will occur, as second-degree or partial thickness burns Marl189are still painful events and may still require medical attention. Furthermore, severe burns can still be experienced on the hands which will usually be closer to a possible arc when testing live circuits than the 450mm given previously.

They are often termed as curable burns however, and the incident energy level, which is the threshold of a second-degree burn, is 1.2 cal/cm² as discussed previously. This is a good starting point for our arc flash strategy and whilst it will not tell us which panel doors contain the tiger, it will tell us where the hazard is less severe.

The calculators and tools that are available with this guide will allow you to quickly determine which circuits are below 1.2 cal/cm² so the first three parts of our risk assessment can be completed for the arc flash hazard. In a generic tiered approach, I have grouped the above calculators and tools as tier 1, 2 and 3. In the following pyramid diagram, Figure 10.1, these groups are listed and at the base are the tier 3 circuits which are the most plentiful in most industrial and commercial facilities. As you can see, the only information that will be required is the type and size of the upstream protection and the minimum prospective fault current level. Tier 2 circuits need actual fault level only and finally, at the top of the pyramid, the actual fault level and a time current curve will be required to return an accurate prediction of arcing current and incident energy. Chapter 14: Hazard and Severity Calculations explains all the tools in detail.

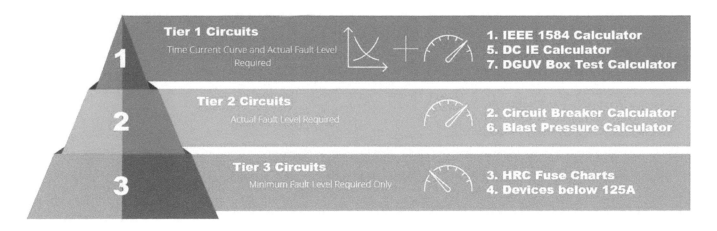

Figure 10.1 Three tier calculator approach

Tier 3 Circuits – Are the low voltage circuits where the upstream device is known but there is no need for a time current curve. All that will be required is the minimum prospective fault current at the equipment to be worked upon. Circuits where the incident energy is below 1.2 cal/cm² can be quickly determined as well as those circuits where there is a dangerous hazard level. These conditions are given in the protective devices that are listed in the calculators and tools that accompany this guide. From experience, you may not know what the actual prospective fault current is, but what is usually apparent is what the fault level is not. For instance, if the named range is 2000 amperes and below, a skilled person would be able to determine the likelihood of that being the actual fault level at the equipment. For other instances, there is a prospective short circuit current calculator which accompanies this guide.

Tier 2 Circuits – Are similar to tier 3 circuits insomuch as they are low voltage where the upstream device is known but there is no need for a time current curve. In this case, what will be required is the actual prospective short circuit current at the equipment to be worked upon. This is providing that the fault level does not exceed the capability of the device.

Furthermore, the protective devices are circuit breakers but are not listed in the calculators that accompany tier 3 circuits detailed above. In the tier 2 there is a circuit breaker calculator for ACBs and MCCBs which will allow you to determine an incident energy level based upon the circuit breaker characteristics. The calculator tool works on the basis of a device instantaneous characteristic and produces an actual incident energy level. From this information parts 4 and 5 of our risk assessment process given previously, can be followed which are; "considering if the risk can be eliminated; and if not, making a judgement on whether further measures to prevent or reduce the risk needs to be introduced".

Tier 1 Circuits - will require details of the time current characteristic curve for the protective device as well as the actual prospective short circuit current at the equipment. The toolbox has an accurate calculator based upon the IEEE 1584 Guide for Performing Arc Flash Hazard Calculations, to determine the incident energy. It also incorporates motor fault level contributions if required.

The above calculators will allow you to recognise extreme danger as part of your risk assessment process and the effort is not too onerous in terms of applying the models. For many service providers it could confirm that for the majority of their operations, the arc flash hazard is not of a high order. It will allow them to apply a rule of exceptions to their safety procedures. In other words, be able to highlight extreme danger which will not be the norm for much of their day-to-day activities. For those service providers who are working routinely on high energy circuits, this will allow a review of their activities to make sure that they are effectively managing the arc flash risk.

There is also a prospective short circuit current calculator and additional tools to plot time current curves for IDMT relays and European style (BS88 and DIN) fuses.

10.2.2 Prevention

Prevention must be the fundamental safety principle for the management of arc flash risk. Please refer to Chapter 5: Prevention which gives a comprehensive view of how the arc flash hazard can be prevented from causing harm. The following highlights those prevention issues which the service provider may need to focus upon.

Dead Working. We must always seek to design out, eliminate or remove the hazard at its source and in most cases, this means dead working. The pressure for working on equipment that is still live is sometimes driven by the client and I give an example of this in my Chapter 12: Myths and Mistakes. Without a doubt, the best method of managing the arc flash hazard is dead working and the service provider must resist the pressure to do otherwise wherever possible.

To work dead, the electricity supply must be isolated in such a way that it cannot be reconnected, or inadvertently become live again, for the duration of the work. As a minimum, this will include the positive identification of all possible supply sources, the opening and locking of suitable isolation points by personal padlocks and for the proving dead at the point of work. Earthing down of current carrying conductors will be a requirement of high voltage installations. As a competent service provider, having full and exclusive control of the process would help but that is not always possible or desirable. For instance, isolation of circuits may introduce other hazards elsewhere in the facility, as an example, circuits in a hospital. Whilst exclusive control of the process may not be possible, the goal must be to have control of the point of isolation or isolations.

We have to remember that the isolation process will necessitate proving that the circuit is dead. Electrical circuits should be considered to be live until proven dead and therefore the hazard could still be present until this process is completed. Therefore, it is recommended that the prediction phase is completed before deciding how to safely prove circuits dead. Where very high incident energy levels are encountered, consider testing at downstream circuits first, using distance and of course PPE for residual risk. There is a large section on proving dead in Chapter 6: Process, Policies and Procedures which, if followed, will reduce the risk and help to eradicate the need for heavy PPE.

10.2.3 Process

The service provider should always have their own rules and supporting procedures and not rely on the client policies in respect of electrical safety. This can sometimes lead to compatibility issues that will have to be dealt with by dialogue between contractor and client. I see the situation as being no different to a commercial contract between the same parties. The contractor would like to supply services to their terms and conditions and clients would like to procure those services under their terms and conditions. The most important point is that safety rules are agreed prior to work commencing. In most cases this can be a simple meeting of minds and an acknowledgement between the two parties of which rules take precedence. In other cases, it may necessitate the client handing over total control of an area or network to the service provider. Examples are assuming control of a high voltage network or working to Construction (Design and Management) Regulations.

Where it is necessary to undertake live work on a client's site or premises, then the risk assessment must give justification and control measures which are detailed in previous chapters and also specifically for service providers below.

Work on or near live conductors needs to be controlled and it is worth highlighting those circumstances where such activities will definitely not be sanctioned by the rules. A non-exhaustive STOP list could be:

- Is there a justification for live working? (See Chapter 6: Process, Policies and Procedures for an explanation) - If the answer is NO then dead working is essential.
- Has the equipment been examined and found to be in good condition? Live work should never be permitted where there are any doubts about the safety of cables and electrical equipment being worked upon or adjacent to those being worked upon.
- Are test instruments, leads and tools approved, calibrated, correct specification for transient over-voltage levels and in good condition?
- Are test leads fused and in good condition?
- Is work at height required? Working at height whilst carrying out live work can create additional problems. Space is restricted and falls create another hazard.
- Is there confidence in the upstream protective device? Has it been adequately maintained?
- Is suitable accompaniment provided? Lone working should not be allowed.
- Are the workers competent and authorised for the task?
- Is the area free from ignition hazards?

Note that there is much more detailed information in Chapter 6: Process, Policies and Procedures.

The above bulleted list is something to consider. It is not exhaustive or mandatory but is an example of how rules can communicate clear boundaries on what is not acceptable practice. I would also suggest adding certain tasks to the list that would not be acceptable. Issues such as connecting and disconnecting cables into live equipment may be one such example unless specifically covered by rigorous safety rules.

10.2.4 Protection

Where the risk cannot be controlled by prevention or where there is a residual risk of injury then it may be necessary to consider mitigation to prevent injury to the worker. The provision of PPE should be used for electrical work where there is risk of injury from all hazards including the arc flash hazard. The requirements for the selection and use of arc rated PPE have been comprehensively covered earlier in this guide, not forgetting other hazards including electric shock.

In Chapter 8: Data Collection, there is a PPE selection model which will work for the service provider involved in inspection work. Otherwise, the 4-tier approach shown in the prediction section above, together with the prevention and policy sections can be used for determining the correct FR rating. As pointed out in the prediction section, severe burns can still be experienced at the hands which will usually be closer than 450mm to a possible arc when testing live circuits. For this reason, insulating gloves with leather over protectors are recommended for all work where the hands may be closer than 450mm, when testing circuits for instance.

10.3 What else can I do to manage the arc flash risk?

One of the most important things that you can do to prevent injury from the arc flash hazard is to provide awareness to any workers who may be involved with energised electrical equipment. From my experience, there are a disproportionately high number of arc flash incidents among contractors. In Chapter 6: Process, Policies and Procedures, there were various factors listed in the scope of technical knowledge or experience when defining competence. Among them was the need to have; an adequate understanding of the system to be worked on and practical experience of that class of system, an understanding of the hazards which may arise during the work and the precautions which need to be taken and an ability to recognise at all times whether it is safe for work to continue. A good risk assessment methodology will assist greatly in providing awareness, information and an understanding of the hazards. To have these principles enshrined in well written safety rules will enhance competence and safe working practices.

Learning Points

- ➢ Contractors appear to be high risk when it comes to arc flash.
- ➢ Do the risk assessment and know whether you are dealing with a raging tiger or a domesticated cat!
- ➢ Use the analytical tools that can be obtained with the guide.
- ➢ Follow the 4Ps principle.
- ➢ Provide awareness of the hazard to your team.

Chapter 11

Electrical Utilities

Because electrical utility companies carry out such a large proportion of their operational work on live equipment, this is justification for a separate chapter to be written to highlight some of the challenges of arc flash hazard management. I produced a report into electrical flashover incident energy for Northern Powergrid in 2010 and the work was seen as ground-breaking by peers in the industry at the time. I believe that the report, and the subsequent approach by Northern Powergrid led to a hugely successful outcome which is something that I would like to share in this chapter.

11.1 So, what makes electrical utilities special?

Live work is routinely carried out on overhead lines and low voltage cables and equipment in a way that is seldom seen on industrial and commercial electrical installations. In fact, the majority of work on low voltage systems is carried out live, and the justification for live working is a balance between the risk to consumers and the risk to employees. Management of the arc flash risk is broadly organised by competence, training, rules and protection.

Operations that are fairly unique to electrical utilities are; a) live low voltage underground cable jointing where skilled employees have to prevent the risk of electric shock and burns in all weathers and often hampered by traffic and the general public, b) low voltage overhead work given the same other hazards as well as work at height by climbing or access platforms, c) link box procedures which often involve manually paralleling supplies from different transformers, d) working on open framed live low voltage feeder pillars and e) high voltage live line work. Some of the above operations are frequently carried out under emergency situations brought on by weather related events. This can mean that employees are working in the teeth of a gale.

Northern Powergrid owns, constructs, maintains, repairs, and replaces assets for electrical distribution across the North East, Yorkshire, and North Lincolnshire region of the United Kingdom to 8 million people across 3.9 million homes and businesses. Its network consists of more than 61,000 substations, has around 30,000 kilometres of overhead lines, almost 67,000 kilometres of underground cables and has over 2000 employees. Northern Powergrid's network has a connected transformer capacity of over 38,000 MVA that operates at voltage ranges from 132 kV down to 240 volts.

11.2 Project Overview

The company was reviewing personal protective equipment (PPE) procurement with a view to rebranded uniforms and commissioned the project in order to assist in the design of clothing systems and arc protection. The objectives were:

1. To determine incident energy levels as part of a risk assessment strategy for a variety of work tasks by putting together a number of scenarios and circuit configurations.
2. Link the above calculations into personal protective equipment requirements to protect against personal injury from electrical flashover.

11.3 Strategy

Working with the Northern Powergrid Field Standards Division, various models were developed to simulate system conditions against site activities and determine the estimated incident energy levels that could be encountered by employees undertaking work on or near energised equipment on the Northern Powergrid system. Various working activities and complex network configurations were examined to create data on which future PPE requirements could be designed and also provide further information for risk assessments. The modelling took place on the low voltage (400 volts) and secondary high voltage (11,000 and 20,000 volt) networks.

11.4 Methodology

The first question was, how to carry out the prediction part of the 4P approach to arc flash risk assessment? It was a hugely complex interconnected network after all, and there was some initial doubt about whether meaningful results could be produced. I was helped by two factors. Firstly, the willingness and cooperation of directors, managers and frontline employees to engage with the project, and secondly, like most distribution networks within the UK, a conservative approach had been adopted towards their construction over the years. This meant that systems, and therefore circuit configuration and protection tended to follow well trusted norms.

Working to the above strategy, various models were created to simulate site activities. This would determine incident energy levels as part of a risk assessment strategy for a variety of work tasks by putting together a number of scenarios, tasks and circuit configurations. The calculations would be useful in informing preventative strategies and indeed produced some surprising information in respect of the effects of fault levels and working distances. This would eventually link the above calculations into Personal Protective Equipment (PPE) requirements to protect against personal injury from electrical flashover. The project was sub divided into three headings as follows:

11.4.1 Low voltage underground operations including:

- Underground cable jointing.
- Metering and service work.
- LV feeder pillars jointing - inspections, live testing, fuse/link replacements, fuse mate applications.
- Link box operations - inspections & testing, linking.
- Industrial service units - inspections, live testing, fuse/link replacements.
- Bulk LV supplies to Industrial Commercial Users, provision of large single feeder ACBs. Possible regeneration effects of large induction motors.
- Application of temporary generation to systems, effects on all the above scenarios.

11.4.2 Low voltage Overhead operations including:

- Work on open 3 phase and neutral distribution lines.
- Shrouding and isolation.
- Tap off operations mains and services.
- Aerial bundled conductors (ABC) terminations.
- Live work by climbing as well as access by mobile electrical work platform.

11.4.3 High voltage Distribution Operations including:

- Racking.
- Testing with voltage indicator (Proving Dead).
- Switching.
- Inspections.
- VT Operations.

The above work groups were designed to be representative of the majority of operations and were not meant to be exhaustive. Primary substations were not included in the calculations because of the high degree of remote operations and voltage restrictions within IEEE 1584 calculations.

Complex risk assessments were undertaken over a four-month period. The risk assessment looked at worse case scenarios, taking photographs of all the high-risk operational tasks and measuring the distance between a potential arc and a worker's body in different positions. The measurements were collated into a report that evaluated the energy levels, arc flash risks and highlighted areas where there was opportunity to improve. The process met European Legislation (European Council Directive 89/391/EEC EU Workplace Health and Safety Directive) requiring employers to perform a risk assessment on tasks and was framed by the 4P approach, See Chapter 3: Risk Assessment and the Four P guide.

The incident energy calculations were carried out in accordance with IEEE 1584 Guide for Performing Arc Flash Hazard Calculations 2002. Two single line diagrams were produced which would give the flexibility to model most circuit configurations at both high voltage and low voltage. Large permanently connected motors and temporary generators were included to add to the complexity of various infeed fault contributions.

11.5 Modelling and Data Gathering

The Northern Powergrid training centre was used as a base for simulated conditions for high voltage and low voltage substation operations. The training centre had most types of switchgear that could be encountered in the field. In addition, the outdoor overhead lines training centre was also used to get actual conditions and body measurements for various overhead tasks. This was all backed up by visits to worksite locations and system design offices to view cable records. Body positioning, working distances, access and egress could all influence severity of injury and were taken into account as part of the assessment. The models that were developed to give circuit configurations included the following:

a) Various cable types, sizes and lengths.
b) Different LV boards.
c) Transformers of varying sizes.
d) Parallel configurations between substations.
e) Temporary generators – contributions.
f) Fault contributions from large LV industrial induction motors.
g) Various protective devices including TLFs, fuses, electromechanical relays and electronic relays.

11.6 Assumptions and Generalisations

A great deal of discussion was had about various assumptions/generalisations that could be made and also which of these could be discounted. The following highlights some of these topics and how decisions were made.

Embedded Generation. Whilst the models simulated various temporary generators, normally installed to keep customers online during maintenance, there was no account taken of smaller embedded generators. It was recognised at that time, that in the future, micro generation may become more prevalent and could have an impact on incident energy levels.

Time Lag Fuses. TLFs were the most popular type of device in use for the protection of secondary distribution transformers and various types were modelled into the scenarios including tagged, barrel type and rewireable. Tagged TLFs were used as the final basis of the studies as they are the most popular and have a greater long-time disconnection as displayed on the time / current characteristic curve. Rewireable fuses were being replaced at the time and were non-existent in some areas.

Circuit breaker opening times. Northern Powergrid operation manuals listed high voltage circuit breaker times as between 40ms for Vacuum up to 75ms for Oil filled equipment. A conservative figure of 83ms was used in the final models. It was assumed that there was a high level of confidence in switchgear operating times and therefore no further safety margin was built in.

Generators. Most generator sets for temporary supplies were hired in and sizes varied from 200kVA to 1200kVA. Information about the generator impedance data was obtained from the main hire company. The models were adjusted for generator sizes which were appropriate for the transformer capacity so an LV generator of 800kVA would be used where the transformer size was 800kVA.

Low Voltage Fuses. The majority of low voltage protection was by BS88 J-Type Links and there were some small variations found between manufacturers. In the end all time/current characteristics were based upon Lawson stated data for total clearance time.

Low Voltage Networks. The vast majority of low voltage circuits within the south of the region was solidly connected with no sectioning of smaller feeders or services. In the north however, it had been the norm in recent years to install downstream fuses for tee offs to achieve grading and automatic disconnection of supplies in no more than 30 seconds. As it happened the models were finally based upon solidly connected LV networks to give conservative outputs.

Secondary High Voltage Distribution Fault Levels. The maximum fault level was taken as 250MVA at 11kV and 350MVA at 20kV. The system models create a number of scenarios in increments from 25MVA up to the maximum fault level. It was experienced that arcing faults on transformer tails and LV Boards would struggle to clear at very low fault levels particularly on 20kV systems.

Two Second Rule. There was a good deal of discussion about long clearance times for protective devices and many engineers would argue that an individual would recoil from an arcing fault of long duration. This will, however, be subject to them having sufficient working space to move away from the arc. In the IEEE 1584 guide there is reference to this which says:

"If the time is longer than two seconds, consider how long a person is likely to remain in the location of the arc flash. It is likely that a person exposed to an arc flash will move away quickly if physically possible and two seconds is a reasonable maximum time for calculations. A person in a bucket truck or a person who has crawled into equipment will need more time to move away."

This is often quoted as the "Two Second Rule" but is really saying that judgement should be used in determining how long a person will stay in a fixed position in relation to an arc. When calculated to the 2002 version of IEEE 1584, incident energy roughly follows an inverse square law so 20cals/cm² at 300mm will become 5cals/cm² at 600mm. However, this relationship between working distance and incident energy is rather more complex with the 2018 version of IEEE 1584 so this will need to be taken into account.

Working space and ability to move away in the event of an arcing fault should therefore be key factors in any risk assessment prior to live work. In the context of UK Legislation this is expressed in regulation 15 of the Electricity at Work Regulations 1989 which states: "For the purposes of enabling injury to be prevented, adequate working space, adequate means of access, and adequate lighting shall be provided at all electrical equipment on which or near which work is being done in circumstances which may give rise to danger."

Furthermore, the HSE guidance on Regulation 15 from HSR 25 Memorandum of Guidance says that: "Where there are dangerous exposed live conductors within reach, the working space dimensions should be adequate to allow people to pull back from the conductors without hazard."

11.7 Results and Conclusions

Low Voltage Operations. Incident energy tables were produced for low voltage boards that were connected directly to a transformer and via low voltage tails. Scenarios were reported for; six separate primary fault levels, four working distances and four transformer sizes. These scenarios were run at 11kV and 20kV. As was expected, the incident energy levels were at their highest in transformer low voltage boxes and associated low voltage boards. This was particularly the case when reduced fault levels were experienced on the primary side of the largest transformers that were modelled. Tables were produced to show the incident energy levels at a typical industrial service unit for various transformer sizes, primary fault levels and working distances on the same basis as the low voltage boards. Industrial Service Units were modelled with a 200kW motor connected and contributing to fault current.

Low Voltage Jointing. The computer model was used to predict the incident energy along a 300mm² aluminium conductor wave form cable. A two second disconnection cut off was factored in once more. Incident energy tables were produced to show the results for a main supplied from a 400 ampere and a 500 ampere, J type BS88 Fuse. The table produced reported incident energy levels at various distances from the substation up to 700 metres for a 400-ampere fuse and 500 metres for a 500-ampere fuse. Also reported were four different working distances.

Public Lighting Cut Outs. Incident Energy levels at public lighting and street furniture cut outs are very often fed directly off a main and the incident energy levels can be assumed to be the same for a three-phase fault will be reduced as they are almost always connected at single phase. The IEEE 1584 calculations are valid only for three phase faults and to use them for single phase events will be very conservative.

Low Voltage Overhead Lines. Three phase and neutral LV open overhead lines are typically aluminium conductors spaced at 300mm to the Electricity Supply Industry Standard. They are usually spaced at 225mm for 3 phase lines and 150mm for single phase lines. Even at the single phase spacing of 150mm, this is at the limit of validity for the IEEE 1584 calculations. The initiation and viability of a low voltage electric arc at such spacing would be difficult to achieve but where cable tap offs are introduced this could be more likely. A spacing of 32mm was therefore, used to simulate a fault in open air.

The majority of LV overhead lines were fed at fuse ratings of BS 88 - 315 amperes or less but in cases where a larger fuse was used, a decision was taken to adopt the maximum incident energy levels reported under LV underground activities.

The vast majority of overhead line work is undertaken from mobile elevated work platforms but there are occasions when energised work is carried out by climbing.

As can be seen from the images in Figure 11.1, the distances to the worker can be greatly reduced and also the linesman can only retract as far as the limits of his belt. The linesman is shown retracted from a possible arcing incident. The working distance for the following are: Lower torso, 450mm; Chest, 600mm; Hands, 450mm; Face, over 1000mm. This meant that reliance cannot be made on a maximum disconnection time when climbing.

Figure 11.1 Live Line Work

The solution may be to carry out individual risk assessments and determine the fault level at the point of work. Charts were developed to display fault level against incident energy for various BS 88 fuse ratings.

Parallel configurations between substations. Link box operations are often perceived to have high risk because a parallel arrangement is being made manually by the operator. Engineers often report arcing usually because of a "bad parallel" between transformers rather than a short circuit at the link box but there is a possibility that arcing can occur due to fault conditions. It was possible to simulate conditions where the incident energy could be in the region of 24 cal/cm^2 with 400A fuses and 28 cal/cm^2 with 500A fuses at the sources feeding from both ends for a working distance of 450mm. In some cases, the engineer may have to stand on top of the link box to carry out operational tasks such as testing or linking. A recommendation was that no live working be permitted whilst a parallel arrangement exists.

Temporary Generator Connections-Fault Contributions. When generators are paralleled onto LV boards this will clearly have an effect on the fault level and incident energy levels on the network. From the data sheets that were created, it was observed that the incident energy could rise from a norm of 20-30 cal/cm^2 to levels over 100 cal/cm^2 in some cases. The protective device at the generator connection and the HV circuit breaker will need to clear the fault independently. Again, a recommendation was that no live working be permitted whilst a parallel arrangement exists.

Fault contributions from large LV industrial induction motors. A 200 kW motor was connected into the model at an industrial service unit and fault contribution was seen to have a considerable effect on incident energy. A 50% increase was observed at the highest fault level. It is unlikely that a motor of that size would be connected directly onto a main feeding a district and that it would normally be fed independently. It can be assumed however, that where such large motors are connected at low voltage then the LV figures given in this report can be raised by a factor of 50%. This information does have significance for system design engineers when dealing with new connection requests.

Incident Energy Levels at Hands and Forearms. For most live working activities, the distance to the hands and forearms is likely to be less than those given to head and torso and therefore, the incident energy is likely to be much greater. This needs to be taken into account when choosing the type of hand protection that will be required to combat arc flash although incident energy tests have been performed on both insulating gloves and leather protectors. It is generally accepted that this combination of hand protection will offer some protection although injury to hands cannot be ruled out. See Chapter 7: Protection for further up to date information.

High Voltage Secondary Substation Operations. A separate "Task and Working Distance Report" for high voltage secondary operations was produced which shows the most common switchgear. Although fully arc protected equipment has been purchased in the previous 15 years, like most Utilities there is a good deal of non-arc protected equipment in use and much of the older switchgear is also oil filled. The IEEE 1584 calculations cannot predict the ballistic effects of disruptive failure inside non arc protected switchgear, nor can it predict the effects of oil ignition. The incident energy levels were reported for those activities where there is interaction with feeder spouts either through racking in, locking off or testing. The data created used a generic arc gap of 153mm and a working distance of 900 mm. Scenarios were created using a 185mm aluminium conductor backed up by standard inverse protection set at 150% and 0.7 seconds at the given maximum fault current. This was all repeated for six different fault levels and four different feeder lengths.

High Voltage (11kV Feeder) cable strikes. A similar exercise to the above was conducted to simulate cable strikes. Incident energy level tables were produced to show the expected levels at damaged 11 kV cables at the same range of fault levels and feeder lengths. There were two data sets, the first shows the incident energy at a working distance of 900mm and the second for 300mm bearing in mind that the employee or third party may be standing over the fault.

Comments on Future Strategy. The data that was produced from the report was primarily for the selection and design of PPE but was also a valuable resource to influence future risk assessments. There were situations that highlighted dangerously high levels of incident energy and this should inform the judgments to allow work on or in the vicinity of energised equipment to be allowed to proceed. The importance of working space and stance when undertaking live proximity work was another learning point. For instance, a fairly modest increase in joint hole size can make a significant difference to incident energy exposure. A further learning point was that that the highest incident energy level on a low voltage main was not at a point nearest to the originating substation, but some 450 metres away. This phenomenon is something that I have referred to as the "fault level paradox" over many years and is talked about in Chapter 4: Predict. For high voltage cables, the opposite was true in the studies and incident energy levels tend to be higher when closer to the supply source.

Clearly, the use of PPE should always be as a last resort and other risk reduction and mitigation techniques should always be explored not forgetting that the first duty should be to remove the hazard entirely. Arcing faults can be predicted and should be part of system design to engineer the problem out. At low voltage for instance, engineers know at which point voltage drop or earth loop impedance may start to become a problem and very often it is system impedance which will affect the severity of an electrical arc.

11.8 Personal Protective Equipment

Armed with the report, Northern Powergrid used the combined expertise of DuPont™ and its value chain partners, J&K Ross (UK distributor) and Dale Techniche (Garment Manufacturer) to provide a solution to the arc flash hazard. Feedback from the operational departments highlighted that the garments worn at the time by Northern Powergrid employees had a poor level of wearer acceptance due to an uncomfortable design. It also highlighted that the garments were not simple or flexible in their use, with many garments not compatible with each other. There was also an uncertainty on how to match the PPE with hazards in relation to arc flash and whether the existing FR clothing provided an adequate Arc Thermal Protective Value (ATPV). In addition to keeping employees safe, having a professional image and a corporate brand was key. The new designs had to not only adhere to the corporate identity of Northern Powergrid, but also had to be adapted to the working environment of the employees so that the highest degree of usability and comfort could be provided.

DuPont™ identified technical fabrics and solutions that not only adhere to safety and corporate requirements, but also protect against the thermal effects of an arc and flash fire. A patented solution was developed with DuPont™ Nomex® on the outside and DuPont™ Kevlar® on the inside to produce a very high thermal barrier against flash fire and arc. Because of the reaction of Nomex® thickening in a flash fire and Kevlar® providing high tensile strength, when such an event occurs, it creates air pockets in the fabric and therefore provides a lightweight flash fire and arc flash solution.

For the Northern Powergrid project two fabrics were worked on: the e+ARC 220RS woven fabric for coveralls and trousers and the A+PANTHER HD 220 RI knitted fabric to make t-shirts and polo shirts. The e+ARC 220RS fabric is developed to specifically protect against the very high level of thermal energy generated by an electric arc. A new concept in the construction of fabrics was used, based on a non-homogenous blend to create an intelligent fabric. This means that when the fabric is exposed to high thermal energy, it responds with an asymmetric shrinkage, generating an extra air chamber to protect the person wearing the garment. The blend offers much higher protection than any market solution, especially when taking into account its low weight, which is ideal for those working at Northern Powergrid

Once the design was confirmed and the fabrics approved, a brief was submitted to DuPont™ manufacturing partner, Dale Techniche. To ensure that only garments of the highest quality were delivered, Dale Techniche matched the e+arc fabric made with Nomex® with state-of-the-art manufacturing processes which were QA controlled by their own employees. During the development process the prototypes were also mannequin tested on DuPont™ Thermo-Man® for four seconds to simulate an industrial fire with temperatures up to 1,000 degrees Celsius to ensure they would be fit for purpose. This life-sized instrumented mannequin is one of the most advanced thermal burn injury evaluation devices in the world, giving the ability to accurately test fabrics and garments under the most extreme conditions. They were also tested on the DuPont™ Arc-Man® test rig in Switzerland to simulate an arc flash, confirming the high protection level of the fabric.

11.8.1 Tailoring and trials

To ensure the sample garments were fit for purpose, Northern Powergrid had a group of around ten employees from each trade group, including linesmen, jointers and fitters, test different designs trial the garments and make suggestions based on their experiences. Dale Techniche and J & K Ross Ltd engaged with both the group of wearers and the project team in order to assess the feedback received once the garments had been worn for a lengthy period of time. This feedback was then used to fine tune the original designs to ensure both optimum ergonomics and user features were incorporated in the final garment designs before being CE certified.

11.8.2 The Finished Package

Fulfilling the project included sizing over 1,600 employees, a process that took a team of four people approximately five weeks, across 14 sites in the North East and Yorkshire. Each employee was individually sized, and the information recorded into a custom database to ensure that when the final kit was received, there would be no doubt that the garments would be fit for purpose. Prior to the process Northern Powergrid identified the specific product mix each employee required in order that they may be protected from their individual risk exposure (based on the tasks they would be expected to perform). As a result, J&K Ross were able to supply them with individual 'employee packs' to protect against electric arc flash and other thermal risks. A combination of shirts, trousers, coveralls, waterproof jackets, salopettes and wind stoppers were provided to each employee in a customised red and black holdall (Northern Powergrid colours) complete with company logo.

The garment packs are based on a layering system that gives different levels of protection depending upon the layers worn for different levels of hazards. Northern Powergrid developed a simplistic PPE matrix that was also issued to all operational employees and clearly conveys what level of protection is required for each activity. The initial quantities supplied were some 3,500 coveralls, 6,000 t-shirts and polo's and 3,200 trousers for employees across the Northern Powergrid sites.

11.9 Protection in practice

Since supplying the PPE, the effectiveness in keeping personnel safe was graphically illustrated in April 2013 when two Northern Powergrid engineers working in a substation on an 11,000-volt cable fault experienced a flashover caused by the disruptive failure of a low voltage switchboard. Despite the intensity of the incident, they were able to make an emergency escape from the substation and thankfully were not injured by the flashover. Both were wearing their full arc flash PPE required for the task that had been carried out in accordance with operational procedures. It was evident that the PPE, effectively a last line of defence in these situations, served its purpose extremely well.

11.10 Acknowledgements

My thanks are extended to the following organisations in providing the material and permissions to enable this arc flash protection project to be put into the public domain.

- Northern Powergrid
- Electrical Safety UK Ltd
- J K Ross
- DuPont™

Learning Points

➢ Create a strategy to target high risk live working interactions.
➢ Follow the 4Ps principle.
➢ The success of the prediction stage requires the willingness and cooperation of directors, managers, and frontline employees to engage with the project.
➢ Involve the workforce at the earliest opportunity.
➢ Design and fit are important to user acceptance.

Chapter 12

Myths & Mistakes

Writing this chapter has allowed me to draw on my own practical experiences to reinforce my belief that electrical engineers, whether involved in design or putting people to work on high-power systems, must consider the effects of arcing. With all the advances in predictive techniques, it is really

not a defence to be unaware of the severity of arcing faults should they occur. We must take t into account not only in designing protection for workers but also for designing protection for electrical systems which are increasingly more complex and subsequently the losses are considerably greater.

12.1 Arc Flash is much more than PPE

The arc flash debate has been polarised around the need for personal protection and, in the UK especially, the aversion to PPE has sometimes deflected attention away from the true losses due to arcing faults. The following is an account of an arc flash that did not cause personal injury and, to my knowledge, did not result in any subsequent serious illness but did result in huge losses which could have been much worse.

One Sunday afternoon in July, some years ago, the Chief Engineer of a large steel rolling mill called me to say that there had been a serious explosion in one of the high voltage substations. I immediately went out there to see if I could be of any assistance. On arrival the subsequent fire was already being suppressed by the fire brigade, so I started to assess the damage. It became apparent to me that the cause of the fire was a large 11,000 Volt capacitor which was insulated by polychlorinated biphenyls (PCB) and was surrounded in turn by mineral insulating oil. Polychlorinated biphenyl was used for many years as an electrical insulator and is now banned because it is a non-biodegradable poison. There had been an electrical flashover inside the capacitor which punched a hole through the inner envelope of liquid PCB and then through the steel tank holding a large quantity of oil. This ignited the oil and surrounding building fabric which severely damaged adjacent switchgear and transformers putting the whole substation out of action. Although the fire had been suppressed, I was concerned about the hazard that the fire brigade would have been exposed to due to the burning of PCB. I knew that PCBs could give off dangerous cancer-causing dioxins and that is why PCB filled equipment could only be disposed of at licensed waste sites where the temperature of incineration was extremely critical. As a result, the fire crews were dispatched to the nearby hospital to receive health checks and tests.

Learning Points

➢ Arc Flash is much more than PPE.
➢ Arcing faults can cause huge losses that are not related to personal protection.
➢ Consider the possible environmental impact.

Unfortunately, the crisis did not end there. The weather had been unseasonal, and it was raining heavily. It was drawn to my attention that the oil together with the PCB was being washed into a nearby lagoon and was forming a slick on top of the water which was only hours away from being discharged into the river Don in Rotherham. Fortunately, I was able to summon a specialist treatment firm to dispatch a tanker to capture the slick and thus averted what would have been a very serious and costly environmental incident. The damage from the incident ran into hundreds of thousands of pounds and that was not counting the production losses. As it happened, my client closed the plant early for the summer holidays which gave me some time to organise emergency repairs to get the plant up and running again.

12.2 "You have to be a maths genius in order to carry out arc flash calculations!"

Somebody said that to me some 14 years ago. At that time, the arc flash calculations were no more complex than the adiabatic equations needed to work out cable dimensions. It is true that the latest version of IEEE 1584 does carry equations that on the face of it do look a little scary as some have up to 13 correction factors. The abundance of correction factors is due to the fact that the equations are empirically derived, in other words they are curve fitting equations from actual arc flash explosions in the laboratory. The fact is, I am no maths genius but have been quite comfortable carrying out longhand calculations for arcing current, incident energy and arc flash boundaries. There is no calculus involved and the height of complexity of the equations probably involves the log button on your calculator, you know, the one that has not been pressed in a while. The engineer in me says that by being able to understand the formula could make sure that I will not be blindly stabbing numbers into software packages and just trusting in the output. The IEEE 1584 Guide for Performing Arc Flash Calculations carries all the calculations and also useful worked examples. Jim Phillips provides online classes to carry out IEEE 1584 arc flash hazard calculations and has produced really simplified spreadsheets which will breakdown the calculations into small chunks and will aid understanding of the process.

Learning Points

➢ A competent electrical engineer can determine arc flash hazard severity.
➢ The tools provided with this guide give the means to determine fault levels and short cuts to provide answers quickly and without the need for complex software.

The simple calculators that are available with this guide will allow you to arrive at an accurate assessment of the severity of the hazard, arc flash hazard boundaries and levels of PPE if required for the task. They also allow the calculation of fault levels and take into account the contributions from induction motors.

12.3 "An Arc Flash Evaluation is a Huge Task"

This is another myth which was levelled at me many years ago and seems to persist. What can be a big task is putting right the basic electrical system information that any self-respecting electrical engineer ought to have at his or her fingertips. A precursor to an arc flash study is to have a single line diagram of the electrical system, a short circuit study and a protection coordination study. Appendix A from IEEE 1584 2002 described all the data that would be required in order to undertake an arc flash study. It broke the data down into four separate categories; the diagram, short circuit study, coordination study and the arc flash study. I have represented the data headings into the pie chart shown in Figure 12.1 and as you can see, the amount of effort required for the arc flash study is less than 10%. That means that the vast majority of information is what many engineers who have responsibilities for an electrical installation will choose to keep. I would certainly feel uncomfortable as a duty holder if I did not have these records in place.

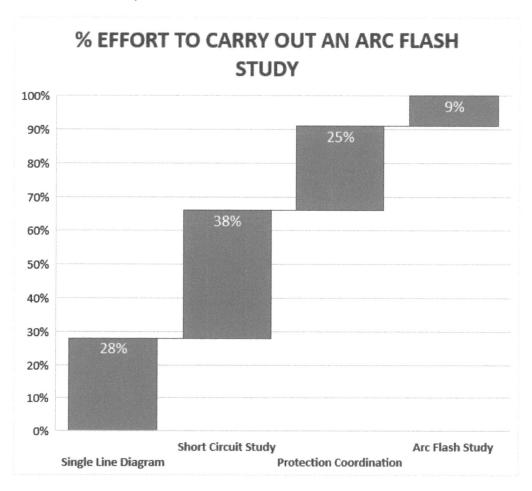

Figure 12.1

12.4 "We do not need to do arc flash evaluation in Europe."

In 1989, what would be known as the "European Framework Directive" number 89/391/EEC (EU Workplace Health and Safety Directive) was passed which introduced measures to encourage improvements to the safety and health of workers. A cornerstone of this directive is risk assessment which means that any employer must evaluate all the risks to the safety and health of workers. For electrical safety, this means that the arc flash hazard cannot be simply ignored even though there are no prescriptive standards in place, such as in the US and Canada and certainly no direct link from the hazard directly into arc flash PPE.

Directive 89/391/EEC creates an obligation on behalf of the employer to assess the level of risk involved in the workplace and the effectiveness of the precautions in place. For electrical work, all hazards should be considered, including the arc flash hazard and not purely shock, as is often the case with many European companies. Arc flash risk assessment for workers who operate in proximity to or on energized electrical equipment, cables and overhead lines, is an essential part of electrical safety management. Electrical work must be carried out with conductors de-energized and isolated wherever possible and there is a very low tolerance for live working in many European countries.

Learning Points

➢ There is an implied requirement in Law to take account of the arc flash hazard.
➢ All hazards should be considered in Risk Assessments.

As a result, many manufacturing plants restrict live working to inspection, diagnostic testing and commissioning purposes. There are, however, many tasks that require working either on, or near to energized equipment. Even then it should also be acknowledged that the process of de-energization often requires exposure to the hazard through interactions such as switching, racking and testing of equipment.

12.5 "You will be killed in the blast above 40 cal/cm^2."

This was somewhat of an urban myth which was perpetuated by the mention in various standards as a threshold for PPE. It has since been debunked and in fact it is possible to purchase PPE of ratings up to 100 cal/cm^2 (maximum rating under IEC 61482-2) because it is technically feasible to do so, and the demand is possibly driven by a failure to reduce PPE ratings by technical means. See Chapter 5: Prevention.

12.6 "Incident Energy below 1.2 cal/cm² means no injury."

Whilst I was undertaking a site survey at a data centre, a pre-qualification document from the main electrical contractor for that site caught my eye. I was told by staff at the data centre that there had not been any recent electrical accidents that had caused injury. But there was a report on page one hundred and something where there had been an injury to one of the electrical contractors' operatives just a couple of years previously. On investigation, it transpired that an electrician had suffered burns to his hand and arm after he had attempted to plug a three-phase miniature circuit breaker onto a live distribution board in the kitchen area.

It was a stupid thing to do as I am sure that the kitchen distribution board was not exactly classed as an essential service. It is what happened next that concerned me. The electrician had presented himself to the local accident and emergency unit and was treated for burns before being discharged that same afternoon with a bandage over his hand and forearm. He returned to work the following day and was put on "light duties". He started to feel unwell, so he returned to the hospital and on presenting himself there he was immediately admitted to hospital for an extended stay. It seemed to me that this "minor injury" had become quite serious and speaking to a good friend of mine, who used to manage a large burns unit, sepsis can be a major complication of electrical burns. I looked at the protection to the distribution board and, if an arc flash study had been undertaken, the incident energy may have been marked up as less than 1.2 cal/cm² at a working distance of 450mm. However, as I have pointed out in Chapter 7: Protection, the hands are often much closer to the arc source and my estimation of the incident energy at hands and forearm was above 13 cal/cm² which can cause serious burns.

> **Learning Points**
>
> ➢ NEVER plug circuit breakers or relays into LIVE electrical equipment.
> ➢ Incident energy at the hands is often of much greater magnitude than the body and face.
> ➢ Burns from arc flash are often difficult to treat.

12.7 Live Working on Damaged Equipment.

I was asked to undertake an investigation into an arc flash incident which caused injury to a facilities technician who had been fault finding on a chiller pump star delta motor control section of an MCC at a large IT manufacturer. The flash over occurred to the integral switch fuse to the panel. You can see the starter panel and the internal damage with the damaged switch fuse removed in figure 12.2. Figure 12.3 shows a healthy unit with 160-ampere HRC fuses.

Figure 12.2 Damaged Starter   **Figure 12.3 Healthy Switch Fuse**

The technician had been called out to the chiller pump and found that it had stopped and the power on indicator at the panel was extinguished. He switched the isolator off at the panel and opened the door to check the fuses, where he found that the 160-ampere red phase fuse had ruptured. He carried out electrical tests to the motor and could not find any faults, so he replaced the faulty fuse. He then closed the panel door and switched the power on at which point the indicator lamp illuminated. He switched the isolator off again and opened the panel door. He then by-passed the door interlocking mechanism on the isolator so that he could visually check that the contactors were activating, switching from star to delta on start up. The electrical flashover occurred after the control panel had been re-energised and the technician had activated the run button for the pump. When pressing the run button, he was holding the panel door slightly ajar, so the resulting incident energy was somewhat shielded from his face and body. He was not wearing any PPE at the time of the incident. Nonetheless, he sustained burns to his hand, arm and face and there was considerable disruption to production. The whole panel was put out of action although the arc was extinguished which was fortunate as the main 1600-ampere incomer to the MCC did not operate. The fault level at the panel was about 30kA but this was increased by motor contribution.

On examination of the switch fuse, I observed that the casing to the unit was severely cracked in several places and this all appeared to be old damage. In addition, the fuses that were fitted into the carriage were all different dimensions. See figure 12.4. None of the fuses would have been factory fitted and I concluded that there had been previous faults on this circuit. The fuse to the blue phase was longer than the originals and it was obvious that it had been force fitted. The casing at its terminal post was cracked and there was arcing on the contact assembly. It was thought that the arcing was confined to the line side of the unit but on close examination there had also been a flashover on the load side. As all three fuses had blown, I concluded that the initial arcing event started on the load side and migrated to the line side of the isolator.

This was all confirmed by a forensic examination by the manufacturer. There was a long list of improvements in the final report but there were a few salient points that I would like to highlight.

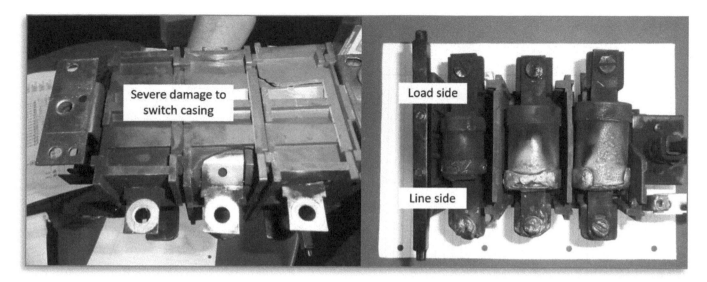

Figure 12.4 Damaged Switch Fuse

Whoever had forced an oversized fuse into the carriage had severely damaged the unit leaving a dangerous defect for one of his colleagues. However, the technician involved in the arc flash was not faultless. There was a risk assessment in place for work on or near live equipment which states that any examination should include looking for signs of damage and not to proceed if observed. There was also a requirement for PPE which was not worn at the time of the incident. There were also several tools and a multimeter inside the panel when the fault occurred and although they did not contribute to the event, it was an indicator of the shoddy way in which the energised work task was approached. There was also a need for a robust sign off procedure for post fault remedial work to large power critical plant.

Learning Points

> Carry out a rigorous risk assessment before working on or near live equipment.
> NEVER work on or near damaged electrical equipment or conductors.
> Beware of the traps left by colleagues.

12.8 Arcing Damage to a Primary Substation

When I was an electrician aged twenty-one, I was to carry out the installation of low voltage transformer tails in a steelworks in Sheffield. Because the work was in a live high voltage substation, I would need to have an access authorisation from the company. It fell to the district engineer to carry out an assessment that would be necessary to issue the authorisation. The familiarisation and assessment involved the district engineer driving me to various substations and pointing out particular hazards and safety features. One of the substations that we visited had recently had a fault which was quite extraordinary in nature, or so I thought at the time. There was a local transformer which fed a low voltage switch board a good distance away using four large single core cables which were bunched together. A fault had developed at the low voltage board and the protection, which was on the high voltage side of the transformer, failed to operate. The result was that the cables burnt for many metres along their length as a result of arcing. The arc was finally extinguished at the point at which the tails went through a wall and all that was left were stumps sticking

> **Learning Points**
>
> ➤ Arc duration is not always measured in milliseconds.
> ➤ Arcing faults are extinguished only if:
>   ○ The source collapses.
>   ○ The protection operates.
>   ○ The arc cannot be sustained.

out of the hole. The substation was a large structure, but everything was covered in black soot. This had a profound effect on me and may explain why I have always been cautious about the issues surrounding the design and protection of transformer low voltage tails. The following story highlights this caution some 35 years after this experience.

12.9 Hospital Design Mistake

I carried out a fault level study, protection coordination study and an arc flash study for a large teaching hospital about 10 years ago. Subsequent work included an upgrade from 6.6KV to 11kV and further protection coordination studies. It was a very old site and much of the electrical installation comprised of legacy switchgear. That said, huge amounts of effort had been put into upgrading electrical supplies on site sometimes as the result of new construction. One major project was a new critical care block. The electrical supplies to the block comprised of 2 MVA transformers and associated low voltage switchboards plus generator backup. I was somewhat surprised to see that a voltage optimisation unit had been installed into the circuit between each transformer and the main switchboards. See Figure 12.5. (Note: Voltage optimisation is a term given to the systematic controlled reduction in the voltages to reduce energy use, power demand and reactive power demand) I was surprised for a number of reasons:

1. Why would someone want to reduce the voltage on a brand-new electrical installation which had presumably been designed properly so that all equipment and lighting would operate at an optimum level?
2. The voltage optimisation units were operating at 400 volts and had no additional short circuit protection apart from that afforded on the high voltage circuit breakers on the primary side of the transformers. This zone is often referred to as a blind spot for protection any fault would have to be seen through the impedance of the transformer windings by the conventional over current high voltage protection relays.
3. I knew that any additional impedance that was inserted into low voltage transformer tails could have a detrimental effect on the disconnection time. This was regardless of any advancement in the understanding of arc flash calculations.

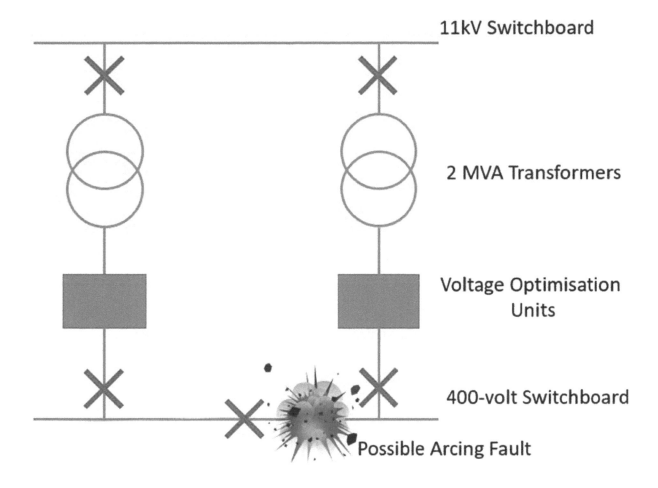

Figure 12.5 Showing location of VOUs

The fact was, that the voltage optimisation units had already been installed and I therefore began to use the nameplate data on the voltage optimisation units in order to model them into my studies. Although an impedance was stamped onto the nameplate I checked with the manufacturer and ascertained that the impedance of the units was dynamic and not fixed. This had a marked effect on the downstream fault level depending upon the settings. Following several discussions with the manufacturers, it was felt that the best way to model the units into the software was to present them as a one to one transformer. In this way, various scenarios could be developed in order to simulate different impedance values. The results were frankly quite startling, and I calculated that the incident energy at the low voltage switchboards could be as much as 73 cal/cm² against 5.1 cal/cm² for not having the units in service. The additional impedance of the voltage optimisation units was like increasing the length of the transformer low voltage tails by a factor of three.

I am afraid that despite receiving contrary impedance values and various assurances about how the internal impedance would act under load, the local healthcare trust was unconvinced, and the units were stripped out. I even enlisted the assistance of a well-respected professor of electrical engineering to help make sense of the restricted technical data given by the manufacturers. The information was withheld because of "commercial sensitivity" so we were effectively dealing blindly with a "black box". The fact was the installation of the units was a very costly mistake but the correct decision to strip them out was made in my view. After all, the units fed the critical care block involving the life or death of very poorly patients.

What I learnt from this experience, was that undertaking the studies had uncovered a design flaw which could have had serious consequences. I would urge anyone undertaking design on large power systems to consider not just zero impedance short circuit faults but arcing faults as well.

> **Learning Points**
>
> ➢ Design Engineers should consider arcing faults on large power systems.
> ➢ Enhanced protection should be considered if necessary.
> ➢ Do not put additional impedance between transformers and switchboards.

12.10 Changing Busbar Tap off Units

From my extensive experience in undertaking electrical surveys across Europe, I came across many unsafe acts, but I would like to highlight three particular examples concerning the changing of busbar tap off units. One was a bakery, the second a famous university and the third was a high-tech manufacturing facility in mainland Europe.

12.10.1 The Bakery

One of the many electrical safety surveys that I carried out was to a bakery. Overhead busbar trunking was used throughout the plant to provide power for machines and associated control panels. One of the two systems in use was nearly 50 years old and obviously long since obsolete. The rating of the busbars was 630 amperes and 800 amperes, and they were protected by HRC fuses. Unswitched plug in fused tap off units were only accessible by a mobile elevated platform or scaffolding and were released by two wing nuts. Engineering staff were uneasy about the practice of releasing the tap off units whilst they were still live but still persisted because of the perceived disruption to the bakery electrical supplies. I had grave concerns for any live work to be undertaken on the busbar system due to the following.

The effect of an electrical flashover occurring with an operative in close proximity is likely to have been extremely serious and resultant burns could be life threatening. This is made worse by the fact that the operative will be constrained by a mobile elevated platform cage or scaffolding. There were signs of water ingress near the 630 ampere busbar trunking and in discussing this with front line staff I found out that there had been a fault as a result. There had also been an electrical flashover at some time in the past on the 800 ampere busbar trunking because I could see the damage from the ground. See Figure 12.6.

Nobody knew the cause of the damage until I spoke with one of the long serving electricians on site who remembered the event vividly. Whilst taking a deep breath, he told me that the flashover occurred whilst one of his colleagues was in fact changing a tap off unit and was hospitalised as a result. The company had no corporate memory of the incident even though all the conditions that caused the accident were still there and waiting to happen again.

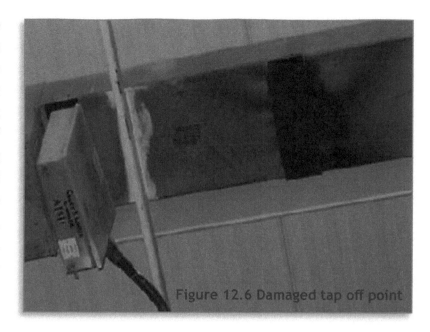

Figure 12.6 Damaged tap off point

12.10.2 The University

I was involved with an electrical safety consultancy for a large university which was spread over several different sites. One of the sites had undergone a good deal of refurbishment work and the electrical distribution system was mainly by busbar trunking with plug on HRC fused tap off units. The backbone of the system was rated at 1350 amperes and appeared to be well designed, in good condition and was a very flexible way of distributing electrical supplies to the campus. There was some discussion with a member of staff over the safety of changing the tap off units whilst the busbars were live as there was a perceived inconvenience to the university in de-energizing the busbars. The way in which the discussion took place unnerved me. He told me that one of his regular contractors had refused to work on the busbar system live and when I asked him what happened next, he told me that they had sacked him. There was apparently, one ongoing experiment at the far end of the busbar that needed security of supplies. He brought in another contractor who willingly agreed to do the work live. He provocatively asked me what I thought about that, at which point I asked him if he had checked with the manufacturer of the system before sanctioning live work. I decided to ring the manufacturer there and then and to my surprise I was put through to one of their directors. With my mobile phone on loudspeaker and placed on my contacts desk, my question was a hypothetical "can I change a tap off unit whilst your 1350 ampere busbar system is still live?" His verbatim answer was, following the sound of air being sucked in through his teeth, "I understand why people would want to do it, but our busbar system is not designed for changing the tap off units live." After the call had ended, I asked the question that if the operation had gone wrong with his second contractor, and injury had been sustained, who did he think would be called on first to give evidence. At this my contact pointed at my mobile phone which was still on his desk.

Figure 12.7 Damaged tap off unit

I discovered later that there has been an incident in the past when, due to water ingress or condensation, a flashover occurred at a tap off position when the unit was operated. I spotted the unit as I was carrying out my inspections and touring the workshop stores. The unit can be seen in Figure 12.7 with arcing damage to the busbar connections.

It may have appeared reasonable at first to change the units live but my view was that the bus bars needed to be made dead and isolated because of the following factors.

• The incident energy level at the tap off units was very high and extremely serious burns could result should an electrical flashover event occur.

• The manufacturer would not recommend it.

• The units were very big and required lifting equipment in many cases.

• Many of the tap off positions could only be reached with access equipment or the access was extremely restricted within service ducts. The inability to vacate the work area safely in the event of an emergency could present a more serious situation.

12.10.3 The European Manufacturing Plant

My third example was a high-tech manufacturing plant in mainland Europe. There were several runs of high-power busbar in the facility, each fed with two 400 ampere parallel feeders. Many machines were fed from these busbars via tap off units and the tap offs were routinely changed whilst the busbar was energised. I did some quick calculations and the incident energy at the tap offs was over 40cal/cm². I noticed that the system was damaged in several places presumably due to being hit by forklift trucks. I was very concerned and asked that energised work be prohibited on this system until further notice. I could not speak the local language and so required a technical translator. I communicated my concerns through the translator to the site electrician. The translator then told me what the electrician had said in reply, which was "does he think that I'm an electrician, or does he think I'm a baker." I found the reply mildly amusing at first but the serious side to this was that his attitude was probably quite common among some maintenance electricians who thought that they were expected to take risks. Perhaps I could have put him back in touch with the bakery in my first example and he could have followed the profession for which he was clearly more suited.

All three examples of busbar interactions have one thing in common. The staff at the locations knew that they were taking some risk but perhaps needed to understand just how much risk was involved. It would be unbelievable that their managers expected them to take risk, otherwise, why would they have gone to the expense of procuring my services. Sounds a bit melodramatic,

> **Learning Points**
>
> ➢ Just because you can plug it in does not mean that it is safe to do it live!
> ➢ Speak to the Manufacturer.
> ➢ DO NOT take risks!

but I have been known to ask a risk taker to put themselves in the shoes of their manager who comes to visit them in hospital after they have put the factory out of action. Is it likely, I ask them, that their manager will say "thank you so much for taking that little risk on my behalf!"

12.11 Finger Safe does not equal Arc Safe

An industry that has accounted for a fair share of electrical accidents over the years is quarrying. I was contracted by one such operation where an arc flash injury occurred after a cable had been connected into a low voltage form 4 construction switchboard. Whilst I was not involved with the investigation, I visited the substation and got a good idea of what had happened.

The connection had been carried out successfully whilst the switchboard had been isolated and locked out. The switchboard was then re-energised, and a labourer was asked to clear all the debris from the surrounding area. The cable was of steel wire armoured and XLPE (Crossed-linked Polythene) 4 core construction so there was a good deal of stripped insulation and of course armour wires left on the floor. The labourer picked up the armours first and one steel wire went through the "finger safe" ventilation panel which was protecting the live incoming terminations to the switchboard. The worker was badly burnt in the resulting arc flash and was psychologically affected to the point that he did not return to work even when his physical injuries had healed.

> **Learning Points**
>
> ➢ DO NOT rely on equipment that is finger safe.
> ➢ Substations and Switch rooms are dangerous places for unqualified staff.

12.12 Competence (or lack of it!)

Over my career, I have seen many examples of where screening, phase shields and internal trunking lids had been missing from control panels. These were clearly a result of engineering staff or contractors not replacing them after repairs, inspections, or additions. An example of this was a compressor mains panel where there were bare 800 ampere live bus bars when the panel doors were opened. The panel was obviously shrouded extensively when new, but this had been removed and not replaced See Fig 12.8. On further investigation I found part of the shrouding had been left on top of an adjacent panel, see Fig 12.9.

Figure 12.8 Shrouding Missing   Figure 12.9 Shrouding Found

The seriousness of this situation was compounded by the conversation that I had with the young man who was working inside that same panel. He was undertaking some diagnostic testing using a multimeter whilst the busbars were energised. He told me that he worked for a refrigeration contractor who had the contract for the fridge plant at that facility. On further questioning it transpired that he had very little experience and the height of his qualifications was a National Vocational Qualification level two in refrigeration. So, the young man was not electrically qualified and working alone on a high-power low voltage control panel that had vital safety equipment missing. Needless to say, I stopped the work immediately and involved the work's manager. No doubt some uncomfortable discussions followed.

Learning Points

⇒ Establish the competence of ALL electrical workers.

⇒ Undertake Panel Inspections and Risk Assessments.

12.13 Connecting Cables into Live Equipment

My role as an electrical safety adviser has involved metaphorically closing stable doors after the horse has bolted on quite a few occasions. By that, I mean that there had been an accident, usually involving electrical flashover that had caused injury and the organisation concerned was being investigated by the Health and Safety Executive. My involvement was to help the organisation to implement safe working practices and rules that should have been in place to prevent the accident. On more than one such occasion, the task which was being performed at the time of the accident, involved connecting cables into energised equipment. What follows are two examples of this practice.

12.13.1 Tyre Manufacturer

All accidents are regrettable and avoidable, but it is particularly poignant when the injured party is a young person. This is the second time that I have come across an apprentice being hurt in an arc flash incident, the first one I spoke about in my introduction to this guide.

An unsupervised apprentice electrician was connecting a cable into the outgoing way of a miniature circuit breaker distribution panel. Most of the distribution panels were mounted at high level on the factory floor so access was gained using a hydraulic lift. A flashover occurred in the panel which persisted because the upstream protection failed to operate. The apprentice received burns and he also injured his ankle as a result of jumping out of the bucket to escape from the arcing inside the panel. The protection did not operate as it was oversized and had not been maintained. The feeder cable was rated at 200 amperes and the protection was 630 amperes. My understanding is that the event that triggered the arc flash event was never established but one theory suggests that there was a defect within the panel which was disturbed by the exertion of performing the connection task. Perhaps a faulty internal busbar connection or maybe a loose item such as a washer or material could have been responsible. In Figure 12.10 it can be seen what the equipment would have looked like on the left and what it looked like after the arc flash event on the right. The flashover appears to be centred on the vertical busbars which would have been insulated.

Figure 12.10 MCB Board before and after

This was a classic case of the domino theory of accident causation. The theory stipulates that injuries occur because of a series of factors, one of which is the accident. In the domino theory, all contributory factors are connected and each one of them is dependent on the preceding factor. It visualises all relevant factors as a series of dominoes standing on edge; so that when one falls, the linkage required for a chain reaction is completed. The final domino is personal injury which would not have happened if any of the other dominos had been removed.

> ## Learning Points
>
> - DO NOT put young people in harm's way.
> - The Dominos:
>   1. Unsafe practice.
>   2. Work at height.
>   3. Equipment manufacturing defect.
>   4. Electrical design flaw.
>   5. Protection failed to operate.
>   6. No personal protection.
>   7. Accident.
>   8. Injury.

12.13.2 The Supermarket

My second example of connecting cables into live equipment came about after a contracting electrician was adding a circuit to a form 4 switchboard. Here again, there had been an accident involving electrical flashover that had caused injury and the organisation concerned was being investigated by the Health and Safety Executive. This time the organisation was a major electrical and mechanical contractor whose client base could be described as blue chip. When I was asked to help implement safe working practices, I quickly realised that there was one other element that could be added to the list of losses from the accident and that was BRAND DAMAGE. It was the first time that I had ever really considered injury to brand image, but in the case of this electrical and mechanical contractor it was definitely critical to them. Having been the managing director of a major electrical contractor, I knew that brand is based on image and image is based upon performance. For most blue-chip clients, safety performance is very important and is often the difference between a successful tender prequalification submission and rejection. Had my client been prosecuted, besides being public knowledge, they would have had to declare a failure of a core competence, that of electrical safety, on any prequalification for work. Definitely not good for business.

The accident happened late one night at a large supermarket after the store had closed. A cable was to be connected into a switchboard presumably when the power could be isolated. However, the switchboard was not isolated, and the plan was that the cable would be connected into a switch fuse in an individual outgoing compartment, see Figure 12.11. Being form 4 construction, the switch fuse was separated from other functional units and main busbars. The incoming side to the switch fuse was still live and insulated. The electrician was very tired having worked the whole day. Inexplicably, he proceeded to remove the insulating barriers from the live side of the switch fuse (See Figure 12.12) at which he caused a short circuit and was injured in the subsequent arc flash.

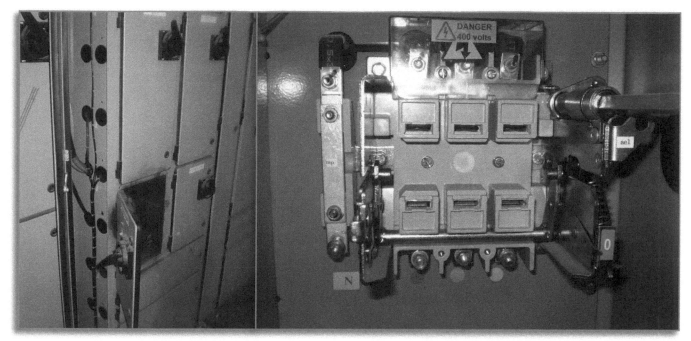

Figure 12.11 Arc damaged board   **Figure 12.12 MCB Healthy switch fuse**

As I said, the electrician must have been very tired at the time and this was probably a contributory factor. Human error contributes to almost all accidents, but it is too easy in circumstances like this to simply dismiss all other possibilities. There are those who would say that it was reasonable to allow the work to proceed given that there was separation of the switch fuse away from all functional units and there was an insulating barrier over the live terminals. Indeed, the UK and Irish regulations for live working states, "No person shall be engaged in any work activity on or so near any live conductor (other than suitably covered with insulating material so as to prevent danger) that danger may arise" etc. However, the words "suitably covered" are dependent upon the task to be performed. Let us look at an extreme example which was an actual accident.

A cable was to be installed into a preformed trench in a substation alongside other cables that were energised at 3.3kV. The other cables were not only fully insulated but were armoured as well. They were encased in cement as they passed from one trench to another, so the idea was to remove the cement to allow the new cable to be pulled in alongside. What was used to perform the task was a jack hammer and you may have guessed by now, the operative punctured one of the adjacent 3.3kV cables. There was a serious explosion and the operative was badly injured. Whilst this could describe other cable strike accidents that occur far too frequently, it illustrates that the words "suitably covered with insulating material so as to prevent danger" are task dependant. Which brings us back to the live switchboard in the Supermarket. The switchboard was form 4 construction, (see Forms of Separation for Low Voltage Switchgear and Control Gear Assemblies to BS EN 61439-2). Some switchboard manufacturers state that form 4 construction is suitable for the connection and disconnection of cables whilst adjacent circuits are live.

However, the BEAMA (British Electrotechnical and Allied Manufacturers' Association) Guide to Forms of Separation, (See Chapter 4: Prediction) points out that specifying a particular form of separation will not guarantee safe working with adjacent equipment energised for any given form number. The BEAMA guide goes onto to say that this means that where live working is being contemplated, a risk assessment and judgement must be made for every situation by the Duty Holder. In this particular case, the control measures relied entirely on the competence of the electrician and implicitly on the design of the equipment. However, there was not a risk assessment in place and/or formal rules and procedures.

> **Learning Points**
>
> ➤ Carry out a risk assessment.
> ➤ Remember the human factors -Fatigue.
> ➤ Do not rely entirely on equipment design to eliminate the hazard.

12.14 PPE – Getting it wrong!

I have come across many organisations that have got it wrong when it comes to PPE. The following is an account of one such situation involving a supplier of construction material and operating a large number of quarries. There was the common mistake of applying a blanket use of PPE with very little evidence of higher order risk controls to prevent the occurrence of arc flash. There was no calculation of incident energy levels or arc flash boundaries which would have given a starting point for the risk assessment process outlined in this guide. Instead, there was a PPE matrix in the company electrical safety policy which had been derived from the local electricity Distribution Network Operator (DNO). A glaring error of interpretation could be seen immediately and that was the belief that the arc flash hazard was much more severe at high voltage than it was at low voltage. In fact, when I did some spot calculations, it became clear that the opposite was true. In addition, the likelihood (and therefore the risk) was less at high voltage than at low voltage. This was because the interactions, switching, with mostly arc protected high voltage equipment were much less likely to lead to injury. This initial modelling indicated that the personal protection for work on high voltage circuits was very much greater than required by quantitative risk assessment and this could have resulted in heat stress to the operatives as well as PPE not being worn.

When I questioned personnel on the comfort of the arc flash PPE currently in use, I found that they were not wearing the currently issued PPE correctly due to the discomfort of wearing the PPE, particularly when the ambient temperatures were high. It was a direct result of the lack of a quantitative and qualitative risk assessment that led to a situation where personnel were too hot in warm weather resulting in excessive perspiration possibly leading to dehydration. There was also restricted movement due to the specification of the PPE. Staff complained of not being able to get in and out of all-terrain vehicles and gloves were often removed in order to work. This all meant that the company electrical safety policy was not being followed and also the cost of PPE was unnecessarily high.

> **Learning Points**
>
> ➤ The arc flash hazard is often greater at LV than it is at HV.
> ➤ Carry out the 4P approach.
> ➤ Remove the need for heavy duty PPE.

12.15 "Am I responsible for the competence of my contractors?"

The fact is that you are responsible. Even if the nature of the work is outside your own competence you will be expected to show due diligence when appointing a contractor. For instance, I remember presenting a strength and capability study to Health and Safety Executive inspectors from the COMAH (Control of Major Accident Hazards) team on behalf of a manufacturing client. The client had subcontracted the operation, maintenance and control of their entire high voltage network to the local Distribution Network Operator (DNO). The DNO had applied their own locks to the high voltage switchgear and transformer tap changers and were operating the system as if they owned it was an extension to their public electricity network. A question was put to my client "how do you know that the DNO is competent?" I must admit that I was a little surprised by the question at first but, having fairly extensive personal knowledge about the outsourcing of high voltage operation maintenance contracts, I could see where the inspector was coming from. Whilst it seemed to be a rhetorical question at first, it clarified that you can subcontract the maintenance but not the responsibility. The fact is, the client owned a high voltage network and it was on their land so anyone entering their site had to be under the appropriate supervision of the duty holder. For this client and others in similar circumstances, I have written specific management of high voltage systems procedures and delivered training to ensure that they fulfil their duty of care to contractors. The procedures were very specific to the site so laid down what, where and when of operational and maintenance responsibilities and activities including emergency situations.

Another example was a prosecution of a food manufacturing plant after the employee of one of their contractors was killed in an electrical accident. Although it was an electrocution case rather than arc flash, it was very clear about the duty holder's responsibilities when it came to contractors from the following press release. "A contractor's employee was removing a redundant cable from a trunking and was killed when he made direct contact with an un-insulated live cable at an in-line connector joint. The subsequent HSE prosecution resulted in a fine of £220,000 and £30,000 costs to the company ordering the work for (i) not ensuring that the subcontractor was sufficiently competent to perform such work and (ii) not ensuring a safe system of work was in place." Other examples are from my experiences where I have, on several occasions found engineers working in energised high-power electrical equipment, with few formal electrical qualifications. One area of concern has been large refrigeration control equipment which often contains high current low voltage bus bar systems. As a result, the incident energy levels are frequently of a high level and could cause severe injury or even death. However, electrical training is purely supplemental to some refrigeration engineers that I have come across and sometimes just a skill that has just been acquired over time.

> ## Learning Points
>
> ➢ You are responsible for the competence of your contractors.
> ➢ Outline the minimum requirements for competence.
> ➢ Ensure that contractors work under your rules including for competence assessment.

Chapter 13

Complex Software Guide

Complex electrical modelling software has improved greatly in recent years and the demand is driven not only by traditional electrical design, but by new technologies such as energy efficiency, smart grids, automation, metering, wind power and energy storage as well as increasing complications with mains disturbances. The result is that you will need to specify exactly which functions you will need to undertake a study as each manufacturer packages their products in different ways. This chapter gives guidance and tips on how to approach the purchasing or leasing of the right package for the purpose of arc flash evaluation.

Going to the trouble and expense of procuring complex software will not guarantee safety nor will it guarantee the accuracy of the result when predicting incident energy. Both these outputs depend upon the following factors:

- That the information that is entered into the software package is correct and has been interpreted properly. Otherwise it is the old adage "garbage in equals garbage out".
- That you have an understanding of the methods of calculation and have sufficient knowledge to question an unusually high or an unusually low result.
- That the results are applied correctly in the context of standards and regulations that exist within Europe.
- That electrical rules and supporting procedures are in place that will adequately and sufficiently give interpretation and understanding of the results to all workers who may be affected.

In broad terms, in order to carry out an arc flash study, you will need to have a package that is capable of providing a single line diagram, a fault level study, a protection coordination study and finally arc flash evaluation. The following tips can be used to make sure that the correct features are in your shopping list especially from a European perspective.

13.1 Compatibility

Software packages are not generally compatible with each other when manufactured by different companies. So, for instance, if a study is being carried out using software from manufacturer A, then all the work that has been carried out in data collection may have to be repeated to model the system in software from Manufacturer B. I have previously highlighted data collection is often a substantial part of the effort required to produce a study so this could be important. Companies sometimes provide data conversion software but from experience, I would want to be convinced that this works effectively before basing any purchasing decision on the availability of such software.

13.2 Single Line Diagram

Most software developers provide evaluation copies of their software, usually up to one month. The evaluation copy will nearly always have a single line diagram and it is worth a test drive to see just how user friendly this is. The single line diagram is after all, the interface that the user will be dealing with the majority of the time and productivity will be driven by how quickly the single line diagram can be built and how components can be entered and edited.

13.3 Fault level study

Most software companies will give a choice of fault level calculations standards. There are two main standards that are used to calculate fault levels in an electrical network. There is the ANSI standard that is used throughout the USA and Canada, and then IEC standard which is used predominantly in Europe. The standards use different methodologies and will arrive at different value in certain circumstances. There used to be a school of thought that, because the IEEE 1584 standard was developed in the USA then the ANSI standard of fault current evaluation was the correct one to use in order to arrive at the bolted short circuit current (prospective fault current). However, the arc flash study is only a small part of the overall process. The strength and capability of the equipment being evaluated must also be taken into account and if it is built and rated in accordance with IEC standards, then the IEC methodology must be used to calculate fault duties.

There is a paper which explores the use of both methods for arc flash studies called "Comprehensive Overview and Comparison of ANSI vs. IEC Short Circuit Calculations: Using IEC Short Circuit Results in IEEE 1584 Arc Flash Calculations" IEEE PCIC Conference proceedings, Cincinnati, OH Sept. 2018. (A. Majd, R. Lou, M. Devadass, J. Phillips). In it, there are sample calculations and comparisons between the two methods. Whilst no generalisations can be drawn about the differences, the paper acknowledges that the IEC method can be effectively used for arc calculations.

13.4 Protection coordination study

In the early days, my data collection efforts did not just consist of discovering site data, but also providing backup to the software developer for missing devices in their libraries. The software libraries for time current characteristics and types of protection were predominantly North American and were limited for IEC devices. This has been broadly overcome, thanks in part to my efforts no doubt, and libraries tend to be much more comprehensive. It is a legitimate question when procuring software to ask about which devices listed and what methods will be used to take account of those devices that are missing. For instance, is there a method of entering time current curve points manually, how does it work and also, what is the response time from the support team for including a missing device?

13.5 Arc flash evaluation

Virtually all software is capable of running IEEE 1584 studies, but this will need checking that the studies are to the latest standard. There are few software companies that have the capability of running the box test arc flash evaluation calculations, but some companies have developments of the standard. This is the German standard DGUV-I 203-078 for arc flash hazard calculations and is recognised in several European countries as well as IEEE 1584-2018. This will allow the user to choose between arc protection class 1 or 2 personal protective equipment in accordance with IEC 61482-1-2 (Box Test) and is described in more detail in Chapter 7: Protection.

13.6 Number of buses

The cost of software is often determined by the number of buses that are required. To purchase software on the number of circuits in a facility is likely to be extremely prohibitive and it is suggested that exclusions of smaller circuits are made for art flash evaluation. If it is considered that every circuit will have to be modelled, then a possible solution may be to breakdown the electrical system into smaller networks and use software with a lower number of buses. It will be prudent to check the cost of upgrading to a higher number of buses as the initial purchase may have been for a bundled offer. There is, therefore, a chance that upgrades may be quite expensive when compared with an up-front cost.

13.7 What else do you need to use the software for?

What other features may be required of the software? If you need to use the software for the evaluation of system disturbances, harmonics motor starting etc it will be good to establish those requirements at this stage. It may be that the software is not capable of undertaking such studies which may mean that such an oversight could be a costly mistake not only in terms of software but also time.

13.8 Labels

Where North American software is proposed, some care is required in selecting the correct labelling formats. As discussed in Chapter 6: Process, Policies and Procedures, the European approach for labelling is less prescriptive than in the USA/Canada but there is a general requirement that employers must provide safety signs where there is residual risk. Warning signs must be standardized across Europe in a way which will minimize the potential problem that may arise from linguistic and cultural differences between workers. U.S style labels based on ANSI Z535 are not acceptable in Europe as the colours and symbols do not conform to the European legislation. The software must be capable of designing European style labels and the language is another consideration. Warning signs for instance, must be yellow (or amber), they must be triangular and the yellow must take up at least 50% of the sign. Information from arc flash studies can be included such as the incident energy level, PPE and various approach boundaries. The table in Chapter 6: Process, Policies and Procedures gives the meaning of each colour.

13.9 Technical Support

The user experience with any computer software is dependent upon the quality and availability of the Technical Support. This is very important when it comes to electrical engineering software support. I know engineers who have chosen to use slightly inferior or more expensive software but where they have a good relationship with the software support technicians. Whilst that relationship might be difficult to measure in the early stages, questions that could give some key performance indicators will be speed of response. So, questions like "how long will it take for a missing protective device to be modelled into the software" may help. Good software companies are proud of their customer facing engineers and their qualifications and experience are usually displayed on their websites, papers and/or training webinars.

13.10 North American Electrical Safety Standards

The Standard for Electrical Safety in the Workplace, NFPA 70E in the USA mandates the use of an Energized Electrical Work Permit. As a result, North American software companies sometimes embed an energised permit printing facility. As detailed in Chapter 6: Process, Policies and Procedures, be careful, as the use of an energized electrical work permit would not be acceptable in the UK as electrical permits are restricted to dead working only.

13.11 Cost

Electrical design and modelling software services is a competitive market, and it is worth shopping around for bundled offers which, will sometimes, present good value for money. Software companies also offer leasing facilities and it is worth the time and effort in order to compare costs over time of software procurement. For instance, there is the option of a perpetual licence but the cost over time will have to take into account the annual support package. Once an allowance has been made for inflation a more attractive proposition may be to lease to software.

The way in which software is packaged varies substantially between manufacturers. This makes it more difficult to make direct comparisons. For instance, manufacturer A could offer a module which includes IEC fault level studies and IEEE 1584 arc flash evaluation as one complete package, whereas manufacturer B will offer them as separate modules. I have also seen very useful additional features, such as direct current arc flash evaluation, thrown in as free features where the same feature may be an expensive separate module from another manufacturer.

Learning Points

➢ Make a shopping list of exactly what functionality you require.
➢ Try before you buy. Most companies provide a free month (with reduced functionality).
➢ The cost is usually driven by the number of nodes or buses.
➢ Make sure that the software will deliver reports, labels, and paperwork to local standards.
➢ Factor in training costs.
➢ Check technical support standards.

Chapter 14

Hazard & Severity Calculators

Introduction

The following chapter maps out the online suite of hazard and severity calculators that are available with this guide as a separate service, (ea-guide.com). The purpose of the calculators and accompanying fuse charts is not to provide complex software calculations of entire electrical networks, but rather to allow the reader to be able to do dynamic risk assessments on a case-by-case basis. The author has had in mind that the person who is responsible for risk assessment is not always in possession of intimate knowledge about the supply network parameters. That is why there is a wide choice of methods available each with a varying degree of input information.

List of Calculators

1. IEEE 1584 Incident Energy Calculator (LEVEL 1).
2. LV Circuit Breaker Calculator (LEVEL 2).
3. Incident Energy HRC Fuse Charts. (LEVEL 2).
4. LV Devices (Fuses and MCBs) up to 125A calculator (LEVEL 3).
5. DC Incident Energy Calculator (LEVEL 1).
6. Blast Pressure Calculator (LEVEL 2).
7. DGUV Box Test Algorithm Calculator (LEVEL 1).
8. Prospective Short Circuit Current Calculator Tool.
9. LV HRC Fuse Time Current Curve Tool.
10. IDMT Time Current Curve Calculator.

Structure of the Calculator Suite

The following diagram, Figure 14.1, shows where the calculators have originated from. The primary calculator and is based upon the IEEE 1584 Guide for Performing Arc Flash Hazard Calculations. In the same group and based on IEEE 1584 there is a low voltage circuit breaker calculator (No 2), IEC 60928-2 HRC fuse charts (No 3) and a calculator for devices below 125 amperes (No 4). Then in group B there are two calculators based on theoretical papers which are DC incident energy (No 5) and a blast pressure calculator (No 6). In group C there is a calculator based upon the German DGUV -I- 203-5188 E Guide to the Selection of Personal Protective Equipment for Electrical Work (No 7). Finally, there are three calculators in group D which provide support in helping to obtain fault current information and also to plot time current curves popular in Europe (Numbers 8,9 and 10).

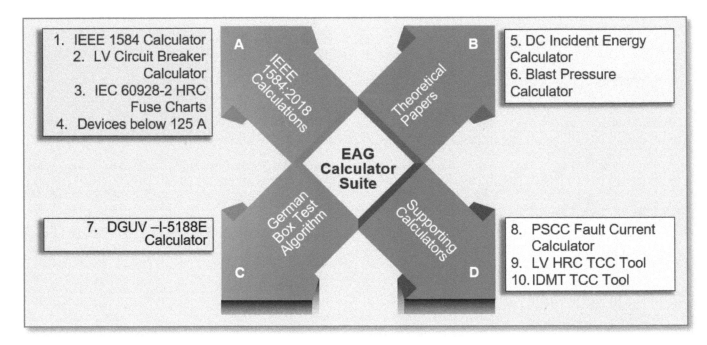

Figure 14.1 Calculator structure

Tiered Approach

In a generic tiered approach, the above calculators and tools are grouped as tier 1, 2 and 3. In the pyramid diagram shown in Figure 14.2, these groups are listed and at the base are the tier 3 circuits which are the most plentiful in most industrial and commercial facilities. As you can see, the only information that will be required is the type and size of the upstream protection and the minimum prospective fault current level. Tier 2 circuits only need actual fault level data and finally, at the top of the pyramid, the actual fault level and a time current curve will be required to return an accurate prediction of arcing current and incident energy. This chapter explains all the tools in detail.

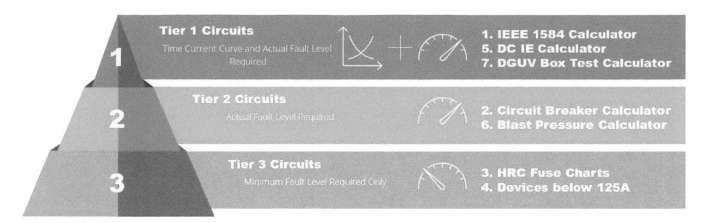

Figure 14.2 Tiered approach

Tier 3 Circuits – Are the low voltage circuits where the upstream device is known but there is no need for a time current curve. All that will be required is the minimum prospective short circuit current at the equipment to be worked upon. Circuits where the incident energy is below 1.2 cal/cm^2 can be quickly determined as well as those circuits where there is a dangerous hazard level. These conditions are given in the protective devices that are listed in the calculators and tools that accompany this guide as a separate service. From experience, you may not know what the actual prospective fault current is, but what is usually apparent is what the fault level is not. For instance, if the named range is 2000 amperes and below, a skilled person would be able to determine the likelihood of that being the actual fault level at the equipment. For other instances, there is a prospective short circuit current calculator available.

Tier 2 Circuits – Are similar to tier 3 circuits insomuch as they are low voltage where the upstream device is known but there is no need for a time current curve. In this case, what will be required is the actual prospective short circuit current at the equipment to be worked upon. This is providing that the fault level does not exceed the capability of the device.

Furthermore, the protective devices are circuit breakers but are not listed in the calculators that accompany tier 3 circuits detailed above. In the tier 2 there is an ACB/MCCB/MCB calculator which will allow you to determine an incident energy level based upon the circuit breaker characteristics. The calculator tool works on the basis of a device instantaneous characteristic and produces an actual incident energy level. From this information our risk assessment process, given previously, can be followed which are: "considering if the risk can be eliminated; and if not, making a judgement on whether further measures to prevent or reduce the risk needs to be introduced".

Tier 1 Circuits - Will require details of the time current characteristic curve for the protective device as well as the actual prospective short circuit current at the equipment. The toolbox has an accurate calculator based up IEEE 1584 Guide for Performing Arc Flash Hazard Calculations, to determine the incident energy for circuits up to 15kV. It also incorporates motor fault level contributions if required.

There is also a DC Incident Energy Calculator which will provide incident energy levels in direct current circuits as part of tier 1.

The above calculators will allow you to recognise extreme danger as part of your risk assessment process and the effort is not too onerous in terms of applying the models.

Also included in tier 1 is the DGUV Box Test Algorithm. Although an incident energy level is not possible using this method, the algorithm is used in some European countries to determine PPE levels. In this case, both a maximum and a minimum fault level will be required as well as a time current curve.

There is also a prospective short circuit current calculator and additional tools to plot time current curves for IDMT relays and European style fuses. The prospective short circuit current calculator also creates maximum and minimum fault levels for use with the DGUV Box Test Algorithm.

14.1 IEEE 1584 Incident Energy Calculator (LEVEL 1)

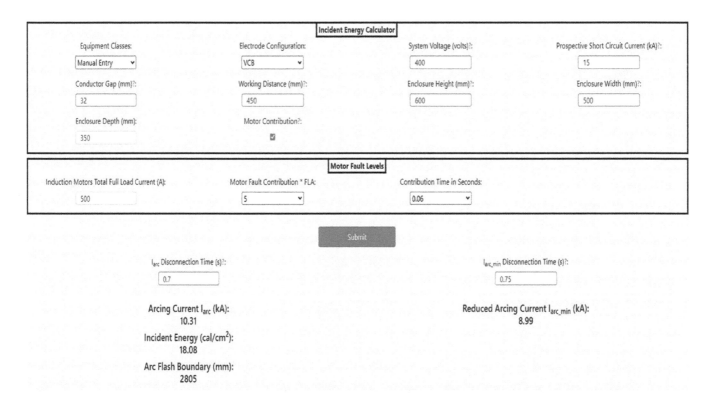

Figure 14.3 IEEE 1584 Calculator

This incident energy calculator is based upon IEEE 1584 2018 Guide for Performing Arc Flash Hazard Calculations. More information can be found in Chapter 4: Prediction. The following sets out how a competent electrical engineer can use the calculator to arrive at an incident energy level at a stated working distance as well as an arc flash boundary which is the distance at which the incident energy level equals 1.2 cal/cm^2 (or the threshold of a second-degree burn).

The IEEE 1584 model is valid for systems having:

- Voltages in the range of 208 V – 15000 V, three-phase.
- Frequencies of 50 or 60 Hz.

- Bolted fault current (Prospective Short Circuit Current) in the range of:
  - 700 A to 106,000 A for voltages between 208 and 600 volts.
  - 200 A to 65,000 A for voltages between 601 and 15,000 volts.
- Working Distances greater than or equal to 305mm.
- Grounding of all types and Ungrounded. (Earthed and Unearthed systems)
- Equipment enclosures of maximum height and width 1244.6mm. The width needs to be larger than four times the gap between the conductors.
- Gaps between conductors in the range of:
  - 6.35 to 76.2mm for voltages between 208 and 600 volts.
  - 19.05 to 254mm for voltages between 601 and 15,000 volts.
- Faults involving three phases only.
- AC (Alternating Current) faults only.

Going through the data field on the form these are described as follows.

Equipment Classes

This is a drop-down field where the user can choose between pre-determined equipment classes or enter bespoke information for conductor gap, working distance and enclosure size. The pre-determined equipment classes are listed as follows.

TYPICAL EQUIPMENT CLASSES DEFAULT VALUES	CONDUCTOR OR BUS GAP	WORKING DISTANCE	TYPICAL ENCLOSURE SIZE	HEIGHT	WIDTH	DEPTH
15kV Switchgear	152	914	LARGE	1143	762	762
15kV MCC	152	914	LARGE	914	914	914
5kV Switchgear	104	914	LARGE	914	914	914
5kV MCC	104	914	LARGE	1143	762	762
Low Voltage Switchgear	32	609	MEDIUM	660	660	660
Low Voltage Switchgear (Shallow)	25	457	SMALL - SHALLOW	508	508	508
Deep LV MCCs and Panel Boards	25	457	SMALL	355	304	203
Cable Junction or Joint box	13	457	SMALL	355	304	203
Cable Junction or Joint box (Shallow)	13	457	SMALL - SHALLOW	355	304	203

Typical Equipment Classes

Selecting any of the typical equipment classes will return the corresponding values of conductor gap, working distance and enclosure height width and depth. It is the user's responsibility to verify that these values are reasonable as they can vary.

If the **Manual Entry** item is chosen, then a second control panel appears called the Manual Entry Table, at which point you can enter bespoke values of conductor gap, working distance and enclosure height width and depth manually. All the units are in mm.

Electrode Configuration

There are five different electrode configurations, and this field is a drop down list giving choices as follows:

- VCB: Vertical conductors/electrodes inside a metal box/enclosure.
- VCBB: Vertical conductors/electrodes terminated in an insulating barrier inside a metal box/enclosure.
- HCB: Horizontal conductors/electrodes inside a metal box/enclosure.
- VOA: Vertical conductors/electrodes in open air.
- HOA: Horizontal conductors/electrodes in open air.

Chapter 8: Data Collection describes which electrode configuration to use as the choice will determine not only the arcing current but also the incident energy & arc flash boundaries. The following diagrams are reproduced here to assist.

IEEE 1584: 2018 Electrode Configurations

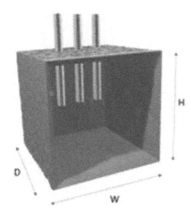

VCB: Vertical conductors/electrodes inside a metal box/enclosure

The VCB configuration uses three equally spaced vertical conductors in a metal box enclosure. The characteristics of this configuration is that the arc is driven away from the source of the supply and downwards to the bottom of the enclosure and out towards the worker. The setup is very similar to the 2002 edition of IEEE 1584.

VCBB: Vertical conductors/electrodes terminated in an insulating barrier inside a metal box/enclosure

The next configuration is VCBB which stands for vertical conductors or electrodes in a metal box enclosure that are terminated into an insulating barrier. What has been demonstrated is that the arc hits the barrier and then the plasma cloud is driven outwards and towards the worker.

HCB: Horizontal conductors/electrodes inside a metal box/enclosure

This is HCB where horizontal conductors or electrodes are pointing outwards inside a metal enclosure. This configuration can be particularly aggressive as the plasma from the arc is directed from the ends of the electrodes and directly at the worker.

VOA: Vertical conductors/electrodes in open air

The final two configurations are in open air. The VOA configuration on the left shows vertical conductors or electrodes in open air. This was also in the 2002 edition of IEEE 1584.

HOA: Horizontal conductors/electrodes in open air

The HOA configurations is similar to VOA but is with horizontal electrodes in open air.

Figure 14.4

System Voltage (volts)

The three-phase line to line voltage in volts.

Prospective Short Circuit Current

This is the user's estimate of prospective short circuit current in kA trying to take into account system variations. The user is advised to enter both upper and lower limits of prospective short circuit current to determine which produces the worst case incident energy. A lower fault current can result in a higher value of incident energy.

Conductor Gap (mm)

The gap between conductors may need some engineering judgement as described in Chapter 8: Data Collection. This is autofilled with the default values from equipment classes that are mentioned previously (or manually if the manual entry option is enabled).

Working Distance (mm)

The distance between the potential arc source (not just the front of the panel) and the face and chest of the worker. If any of the above equipment classes are chosen, this box will be auto filled from table 14.4 above.

Enclosure Height (mm)

Enter the height of the enclosure in mm. If any of the above equipment classes are chosen, this box will be auto filled from table 14.4 above. If electrode configurations VOA or HOA (arc in air) are selected previously, then this box will not be applicable.

Enclosure Width (mm)

Enter the width of the enclosure in mm. If any of the above equipment classes are chosen, this box will be auto filled from table 14.4 above. If electrode configurations VOA or HOA (arc in air) are selected previously, then this box will not be applicable.

Enclosure Depth (mm)

Enter the height of the enclosure in mm. If any of the above equipment classes are chosen, this box will be auto filled from table 14.4 above. If electrode configurations VOA or HOA (arc in air) are selected previously, then this box will not be applicable.

Submit Button

Press the submit button after the above data has been entered and this will return an arcing current I_{arc} and a reduced arcing current I_{arc_min} providing there are no invalid entries.

I_{arc} Disconnection Time

A value of time in seconds needs to be entered for the corresponding Arcing Current (I_{arc}). This will be from the upstream protection device time current curve. As shown below.

I_{arc_min} Disconnection Time

A value of time in seconds needs to be entered for the corresponding reduced arcing current (I_{arc_min}). This will be from the upstream protection device time current curve. As shown below.

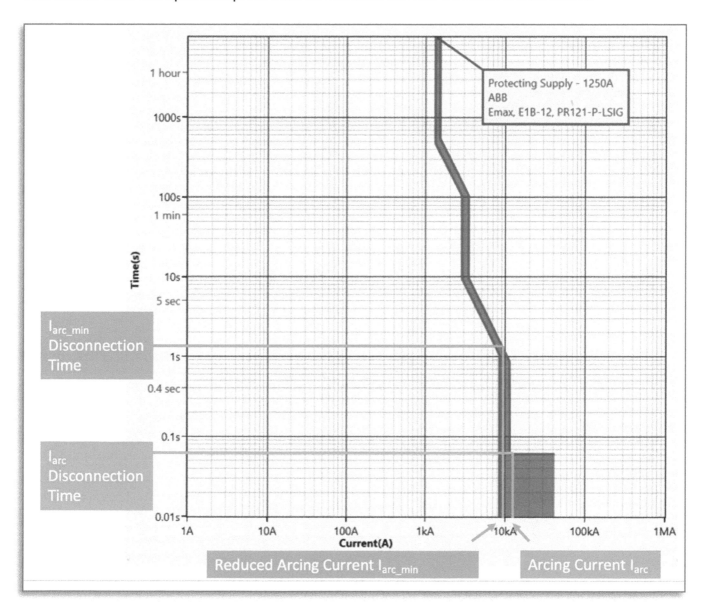

Figure 14.5 Arcing Current Variations

The example in Figure 14.5 is for a typical low voltage air circuit breaker and as can be seen the arcing current in this case is 12.2 kA (marked in blue) with a corresponding I_{arc} disconnection time is 0.06 seconds. The reduced arcing current Iarc_min is 9.9 kA (marked in green) with a corresponding I_{arc_min} disconnection time of 1.2 seconds.

Add Motor Contribution

This is a tick box if you wish to add motor contribution. If there is significant contribution from induction motors, this will have the effect of increasing the prospective short circuit current for a finite time and therefore increasing the incident energy level. As can be seen in the diagram Figure 14.6 below, an amount of incident energy from the motor, which is represented by the pink box, is added to the calculated incident energy contribution from the utility.

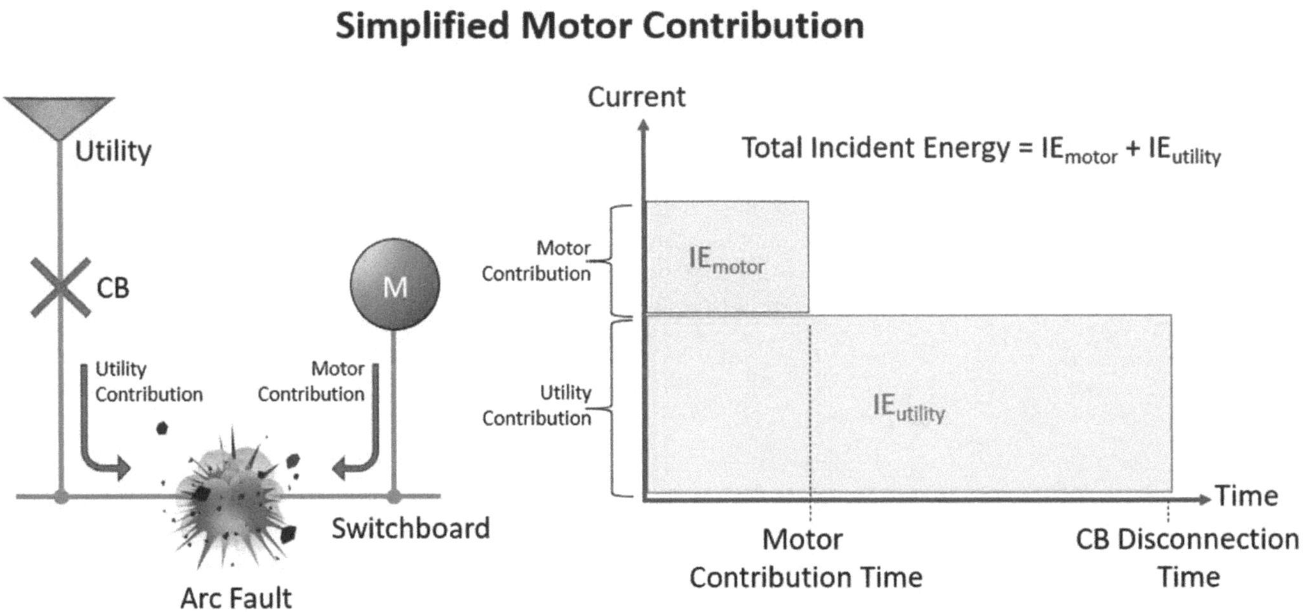

Simplified Motor Contribution

Figure 14.6 Motor Contribution

By ticking the **Add Motor Contribution** box, a separate control panel will be shown with the following fields.

Induction Motors Total Full Load Current

This field asks for the induction motor full load current in amperes which can be either a single or grouped motor load in amperes.

Motor Fault Contribution * FLA

This field is motor fault contribution times full load current which is an engineering judgement. A motor contribution of between four- and six-times full load current is usually acceptable.

Contribution Time in Seconds.

The final field is a drop-down field for the motor contribution time in seconds. Again, this is to be based upon engineering judgement, but induction motor contribution typically lasts from one to four cycles (0.02 to 0.08 seconds at 50Hz) and synchronous motors, this can last from six to eight cycles (0.12 to 0.16 seconds). In summary, the IEEE 1584 – 2018 Calculator follows the logic in the flowchart given in Figure 14.7.

For the sake of simplicity, the calculator shows motors connected downstream of the fault meaning that the sum of utility fault current plus motor fault current will be applied for whatever time that is entered into the motor contribution time field. This may give a slightly higher incident energy in cases where the protective device operates quicker than the motor contribution time. Motors that are connected upstream of the protective device need to be factored into the mains contribution which is derived from a separate fault level study as this may impact on the protective device operating time.

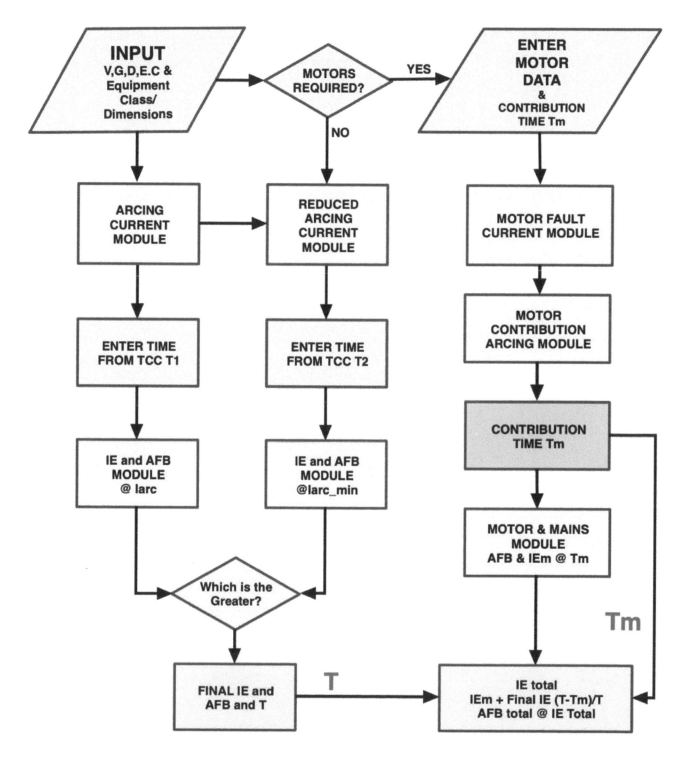

Figure 14.7 Calculations flowchart

Further Important Information

All fields that have specific validation are accompanied with an instructional tooltip, denoted by a "?". If the instruction is followed correctly, in most cases the field will appear green, if it is erroneous, it will appear red.

Results.

The results are displayed in two formats: Format 1 where the arcing current I_{arc} and associated disconnection time returns the highest incident energy and arc flash boundary. Format 2 displays the incident energy and arc flash boundary for the arcing current I_{arc_min} where this is higher than format 1. This is usually because of longer disconnection times due to the lower arcing current.

14.2 LV Circuit Breaker Calculator (LEVEL 2)

The Circuit Breaker calculator tool follows the principle that a calculation is made to determine the minimum fault current to operate a circuit breaker instantaneously. It will allow for results from low voltage air circuit breakers (ACB), moulded case circuit breakers (MCCB) and miniature circuit breakers (MCB).

The calculator has several advantages as follows.

- The user will be able to check against a list of circuit breakers in common use in Europe.
- The user will require an estimate of the fault current at the equipment and does not require a copy of the time current curve providing the upstream device is listed on the drop-down menu or the upstream device meets with the criteria detailed below.
- This is a major advantage of the guide as it fits with the philosophy of simplicity.
- The method will help to establish real danger where settings are borderline.

The following graph shows the Time Current Characteristic for a typical LV Circuit Breaker

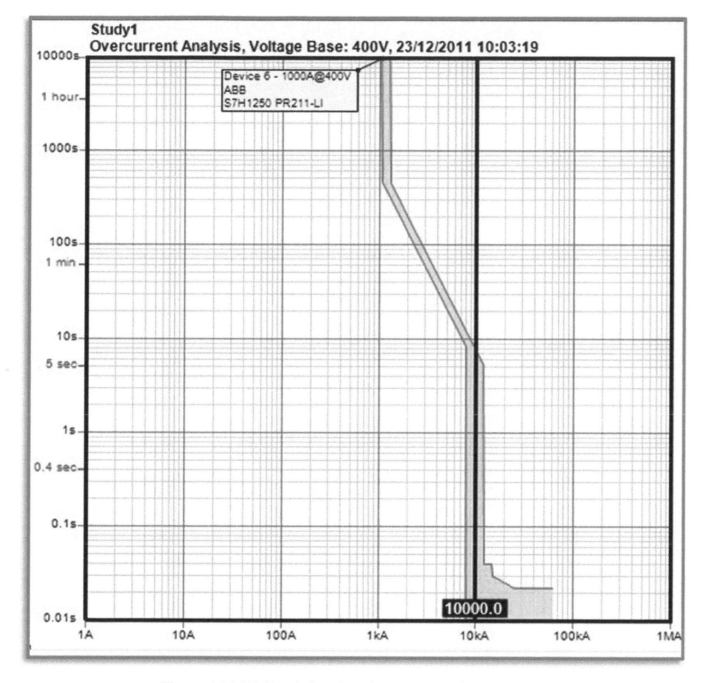

Figure 14.8 LV Circuit breaker time current characteristic

As can be seen from the typical example given in Figure 14.8, the instantaneous pick-up setting from the circuit breaker is shown. In this particular case, this is 10 times the CB rating of 1000 amperes which we will call (I_m). Time current curves from major manufacturers such as Siemens, Mitsubishi, Terasaki, Schneider, ABB, Eaton, Dorman Smith are mostly within a maximum tolerance of 25% but the individual circuit breakers are listed to state that they have been checked. Also shown is the minimum arcing current to give instantaneous operation of the CB and the listed circuit breakers all fall within a maximum instantaneous clearance time of 80ms. This figure includes allowances for curve tolerance and for the arcing current variation correction factor (I_{arc_min}) from IEEE 1584 2018 guide. Where a circuit breaker is not listed, the user can verify that the CB fulfils the clearance criteria of 25% tolerance and minimum instantaneous operating time of 80ms.

The calculator determines the arcing fault current from the data that the user has provided which will be the electrode configuration, conductor gap, voltage, instantaneous setting and prospective short circuit current. The calculator has a YES/NO drop-down menu to state that the instantaneous element is switched on. The incident energy and arc flash boundary will be displayed after the user has entered the working distance and be based on the minimum operation time for the circuit breaker (I_t) and arcing current (I_{arc_min}). The drop-down menu will be displayed which will list all the circuit breakers to which the calculator applies. Should the arcing current be below the current to give instantaneous operation of the ACB the incident energy will be calculated based upon a maximum time of two seconds*. The following diagram, Figure 14.9 shows a typical over current characteristic.

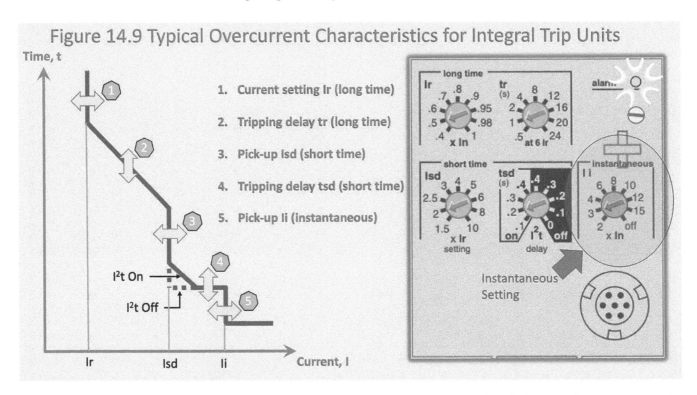

Figure 14.9 Typical Overcurrent Characteristics for Integral Trip Units

1. Current setting Ir (long time)
2. Tripping delay tr (long time)
3. Pick-up Isd (short time)
4. Tripping delay tsd (short time)
5. Pick-up Ii (instantaneous)

*NOTE: IEEE1584 Guide 2018 says "If the total protective device clearing time is longer than two seconds, consider how long a person is likely to remain in the location of the arc flash. It is likely that a person exposed to an arc flash will move away quickly if it is physically possible, and two seconds usually is a reasonable assumption for the arc duration to determine the incident energy."

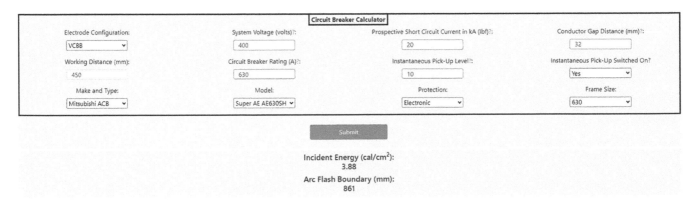

Figure 14.8 LV Circuit breaker calculator

The following is a description of the input/output fields:

Electrode configuration

There are five different electrode configurations, and this field is a drop-down list giving choices as follows:

- VCB: Vertical conductors/electrodes inside a metal box/enclosure.
- VCBB: Vertical conductors/electrodes terminated in an insulating barrier inside a metal box/ enclosure.
- HCB: Horizontal conductors/electrodes inside a metal box/enclosure.
- VOA: Vertical conductors/electrodes in open air.
- HOA: Horizontal conductors/electrodes in open air.

Chapter 8: Data Collection describes which electrode configuration to use as the choice will determine not only the arcing current but also the incident energy and arc flash boundaries.

System voltage

The three-phase line to line voltage in volts.

Prospective short circuit current (kA)

This is the user's estimate of prospective short circuit current in kA whilst seeking to consider system variations. The user is advised to enter both upper and lower limits of prospective short circuit current to determine which produces the worst-case incident energy. A lower fault current can result in a higher value of incident energy.

Conductor gap distance (mm)

Enter the gap between conductors at the equipment. This may need some engineering judgement as described in Chapter 8: Data Collection.

Working Distance (mm)

This is the distance between the potential arc source (not just the front of the panel) and the face and chest of the worker.

Circuit breaker rating (amperes)

This is the overcurrent rating of the circuit breaker or long-time current setting as shown in Figure 14.9 earlier, (not the frame size or rating).

Instantaneous pickup level

This is the instantaneous setting which is usually between 2 and 15 times the overcurrent circuit breaker rating for integral electronic trip units. This may be a fixed value on older thermal magnetic trips and miniature circuit breakers.

Instantaneous pickup switched on?

Verify that the instantaneous section or setting is switched to on. The calculator has a YES/NO drop-down menu to state that the instantaneous element is switched on by selecting YES. Select NO if the instantaneous setting is switched off.

Make and type

Pick from a drop-down list of circuit breaker make and types. This is the first check to see if the circuit breaker is listed in a library of circuit breakers in use in Europe. For the circuit breakers to be listed, they will pass the criteria given later under circuit breaker listings. You can also select "(Valid) Make/Type Not Listed" in which case the circuit breaker meets the criteria but is not listed or "(Invalid) Make/Type Not Listed" and in this case the circuit breaker will always default to two seconds disconnection time.

Model

Pick from a drop-down list of circuit breaker models.

Protection

Pick from a drop-down list of circuit breaker integral protection unit types if fitted.

Frame size

Pick from a drop-down list of circuit breaker frame sizes if given.

Incident energy (cal/cm²)

The calculator will produce an incident energy level based upon either, instantaneous operation of the circuit breaker or a two second maximum disconnection time.

Arc flash boundary (mm)

The arc flash boundary will also be stated based upon either, instantaneous operation of the circuit breaker or a two second maximum disconnection time.

Operating Logic

The calculator flowchart, Figure 14.11, shows the input sequence and the operating logic.

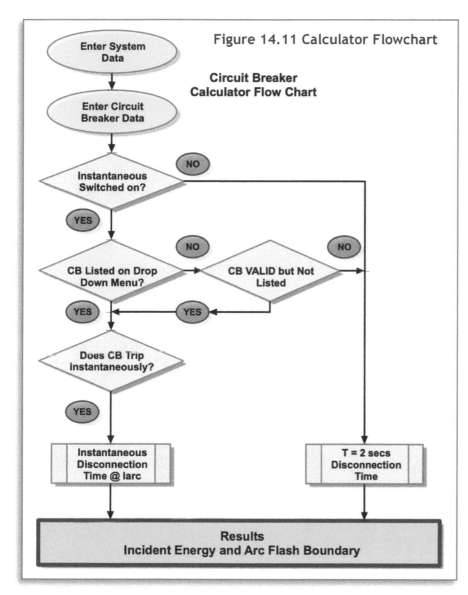

Figure 14.11 Calculator Flowchart

Circuit Breaker Calculator Flow Chart

To use the calculator, first examine the upstream circuit breaker settings and determine the prospective short circuit current at the equipment to worked on or nearby. Verify that the circuit breaker is listed or that it complies with the tolerance values given in the drop-down field "Circuit Breaker".

Enter the working distance (mm) which is the distance from an arcing source to the torso and/or head.

Determine the electrode configuration and conductor gap distance (mm), more information on electrode configuration is given in Chapter 8: Data Collection.

Enter the circuit breaker rating in amperes and the instantaneous pick-up level which has to be given as a multiplier of the rating. Verify that the instantaneous section or setting is switched to on and select YES from the drop-down menu. Select NO if the instantaneous setting is switched off.

A typical circuit breaker integral trip unit is shown in Figure 14.10. From the information given the calculator will produce an incident energy level and arc flash boundary based upon either, instantaneous operation of the circuit breaker or a two second maximum disconnection time. IEEE1584 Guide 2018 says "If the total protective device clearing time is longer than two seconds, consider how long a person is likely to remain in the location of the arc flash. It is likely that a person exposed to an arc flash will move away quickly if it is physically possible, and two seconds usually is a reasonable assumption for the arc duration to determine the incident energy."

Note: you must ensure that the short circuit rating and maintenance of the circuit breaker is correct and fit for purpose.

Circuit Breaker Listings

The circuit breaker listing library contains over 8000 common circuit breakers in use in Europe. In order for the circuit breakers to be listed, they will pass the following criteria.

- The instantaneous element will be adjustable.
- The clearance curve tolerance will be no greater than 25% as explained previously.
- The minimum operation time for the circuit breaker will be no more than 0.08 seconds when operating in the instantaneous zone.

If the circuit breaker is not listed, then you can check that the parameters fit the above criteria and click the checked box. If there is no adjustable setting for the instantaneous setting, then check the manufacturer's data.

The other thing to look out for is some circuit breakers provide a selective short circuit function with a time delay setting as in this example given in Figure 14.12. Make sure that the time delay is disabled and, in this case, t_{sc} should be set to INST. As can be seen in this example, the selective short circuit setting can be set to instantaneous in which case there will be no intentional delay or, alternatively, a delay can be introduced.

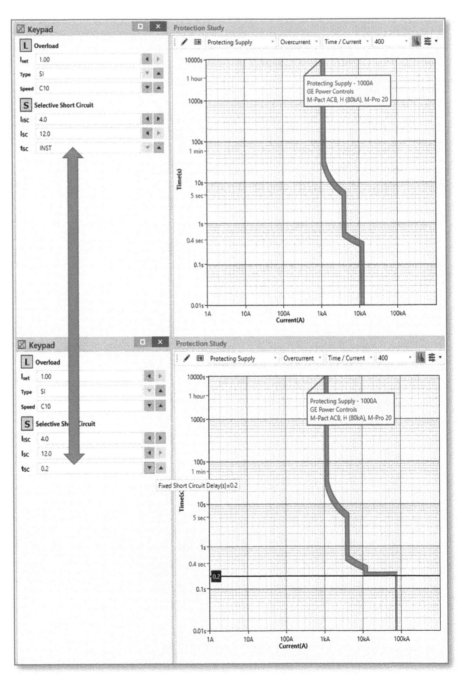

Figure 14.12 Example of time delay setting

14.3 Incident Energy HRC Fuse Charts. (LEVEL 2)

The fuse charts give a very quick and simple way of determining incident energy as part of dynamic risk assessments as well as exposing danger. They are based upon the two most popular fuses in Europe which are the BS 88 and DIN NH models. Both are extremely efficient in clearing high energy faults very quickly and safely where there is sufficient available fault level. The range of the fuse charts are as follows.

- The voltage used is 400 volts which is the harmonised voltage in Europe.
- Fuse sizes are from 160 amperes up to 800 amperes.
- The charts are valid for IEEE 1584: 2018 Guide for Performing Arc Flash Calculations.
- Electrode configurations covered are VCB and VCBB. Other configurations will need to be determined using the IEEE 1584 calculator tool.
- There is a two second maximum reaction time* applied which means that sufficient space is available to pull back from an arcing event. So, the charts are not valid where the worker is confined by the constraints of the working area.

*NOTE: IEEE1584 Guide 2018 says "If the total protective device clearing time is longer than two seconds, consider how long a person is likely to remain in the location of the arc flash. It is likely that a person exposed to an arc flash will move away quickly if it is physically possible, and two seconds usually is a reasonable assumption for the arc duration to determine the incident energy."

The principle behind the charts is that the area which is outside the current limiting zone of the fuses, is expressed as the danger zone. The following chart, Figure 14.13, shows a plot of incident energy against prospective short circuit current at the equipment in question. This is derived from a calculation of arcing current and the corresponding determination of disconnection time provides the basis for incident energy levels to be established over the fault handling capability range of the fuse. This is based upon a working distance of 450mm between the source of the arc and the worker.

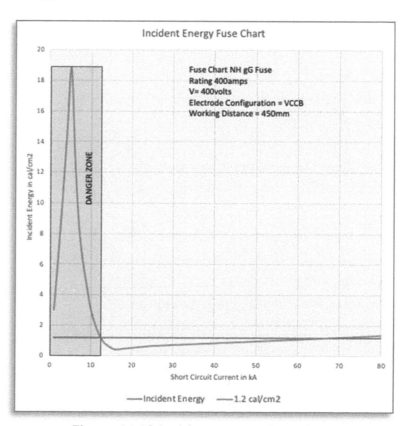

Figure 14.13 Incident energy fuse chart

From the above, we can highlight the key parameters of the danger zone and the fault current at which the corresponding arcing current that will cause the fuse to operate the current limiting zone. In this case, the 400 ampere fuse will protect in the current limiting area for prospective short circuit current values greater than 12 kA. Below that figure, operation will be slower meaning that the incident energy level will build up to 19 cal/cm² at which point the maximum disconnection time of two seconds will kick in. For simplicity, the charts will show the zones of protection as follows in figure 14.14. In this case a BS 88 315 ampere fuse link is used.

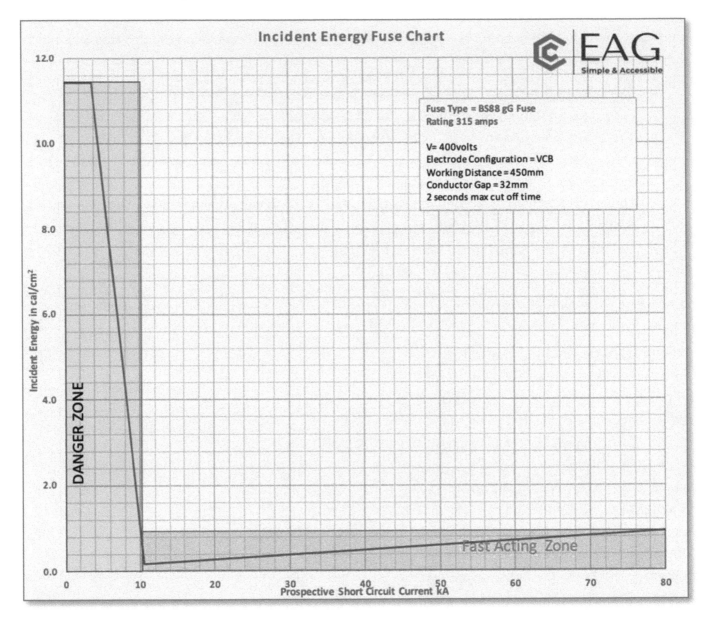

Figure 14.14 Simplified IE fuse chart – for illustrative purposes only

All that the user will require is an estimate of the prospective short circuit current at the equipment and to check that the chart will fall in line with the range limitations given previously. Use of the charts will give an instant prompt of whether to STOP or to carry out further action to ensure the safety of workers.

14.4 LV Devices (Fuses and MCBs) up to 125A calculator (LEVEL 3)

Most fuses and circuit breakers that are 125 amperes or less in Europe provide protection against arcing faults, but care is required as there can be fairly high levels of incident energy even below ratings of 125 amperes. As has been expressed previously throughout this guide, it is usually because of high source impedance meaning that there is not enough prospective short circuit current to operate the protective device instantaneously. In these cases, the device sees the fault as being an overload which will take time to clear and as incident energy is directly proportional to time, there can be a high release of thermal energy in an arc before the fault is automatically disconnected from the supply. However, in most cases, the types of devices that are commonly used in final circuits can disconnect safely and without damage providing that there is enough prospective short circuit current (PSCC) to operate the protective device instantaneously. This calculator allows you to determine the incident energy for when the listed devices are operating in the instantaneous region and also where the device operates outside this region. You do not need to have a time current curve or exact knowledge of the fault current providing that you know that it is not below the instantaneous threshold. This gives an option to manage by exception and be able to discount smaller circuits where they fit the criteria. This allows more focus to be made on higher power circuits which often make up only 10 to 20% of circuits on typical industrial and commercial systems.

This incident energy calculator is based upon IEEE 1584 2018 Guide for Performing Arc Flash Hazard Calculations. More information can be found in Chapter 4: Prediction. There are listings for all five electrode configurations and also working distances of between 305mm and 900mm. The voltage levels are for 400 volts harmonised in Europe and are for three phase symmetrical arcing faults only. The screen shot of the actual calculator is shown in Figure 14.15. The devices are the most common types in use which include the DIN NH and BS 88 types of high rupturing capacity fuses and miniature circuit breaker to EN 60898 and EN 61009-2.

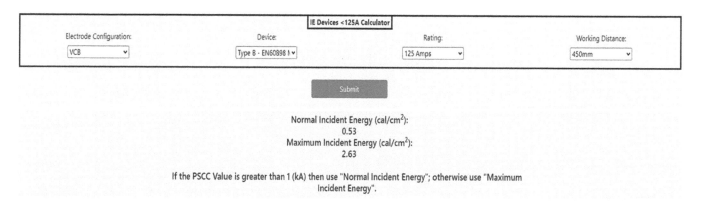

Figure 14.15 IE Devices <125A

The fields on the data entry and output form are as follows.

Electrode Configuration

There are five different electrode configurations, and this field is a drop-down list giving choices as follows.

- VCB: Vertical conductors/electrodes inside a metal box/enclosure.
- VCBB: Vertical conductors/electrodes terminated in an insulating barrier inside a metal box/enclosure.
- HCB: Horizontal conductors/electrodes inside a metal box/enclosure.
- VOA: Vertical conductors/electrodes in open air.
- HOA: Horizontal conductors/electrodes in open air.

Chapter 8: Data Collection describes which electrode configuration to use as the choice will determine not only the arcing current but also the incident energy. An assumption of 508 x 508 x 508mm enclosure size and an arc gap of 32mm has been made.

Device

A drop-down list menu gives a choice between DIN NH and BS 88 types of high rupturing capacity fuses and three phase miniature circuit breakers to EN 60898 and EN 61009-2.

Rating

A drop-down list menu gives a choice between the range of 32 amperes up to 125 amperes.

Working Distance

A drop-down list menu gives a choice between 305mm, 450mm, 600mm and 900mm from a possible arcing source and the torso and head. Be aware that the hands can be closer to an arcing source.

Normal Incident Energy in cal/cm² (CALCULATED)

This gives a value of incident energy for the configuration, working distance and rating chosen above where the PSCC is high enough to allow instantaneous operation of the device.

Maximum Incident Energy (CALCULATED)

This gives a value of incident energy for the configuration, working distance and rating chosen at or below the value of PSCC that would be required to allow instantaneous operation of the device. A two second disconnection time* has been applied.

*NOTE: IEEE1584 Guide 2018 says "If the total protective device clearing time is longer than two seconds, consider how long a person is likely to remain in the location of the arc flash. It is likely that a person exposed to an arc flash will move away quickly if it is physically possible, and two seconds usually is a reasonable assumption for the arc duration to determine the incident energy."

If the PSCC value is greater than XX (kA) then use "Normal Incident Energy" otherwise use "Maximum Incident Energy" (CALCULATED)

The value shown by "XX (kA)" is the prospective short circuit current above which will operate the protective device instantaneously. If the PSCC is equal or less than this given value, then use the value given by "Normal Incident Energy" otherwise use "Maximum Incident Energy".

Conditions and Assumptions

You are advised to check that these are reasonable assumptions:

- That the device capacity is not exceeded by the maximum value of PSCC, (strength and capability).
- Based on clearance curves generic device types from Amtech software.
- 508 x 508 x 508mm enclosure size.
- 32mm gap between conductors.
- No motor contribution considered.
- The reported incident energy figures are found by rounding up in 0.25kA segments.
- 400 volts systems only.
- Instantaneous trip times considered are a full cycle 20ms for MCBs and a half cycle of 10ms for fuses at 50 Hertz.
- Highest PSCC considered is 15 kA for MCBs and 80 kA for fuses.

14.5 DC Incident Energy Calculator (LEVEL 1)

Introduction

There have been several papers that have addressed the calculation of incident energy in direct current systems. Up to 2021, no standards that have the same standing as IEEE 1584, but it is likely that this will be addressed in coming years. That is of course, driven by the huge demand for calculations as DC systems become more abundant because of advances in solar and storage technology. Jim Phillips carried out research into the various methods and in 2010, introduced DC arc flash calculations into his Arc Flash courses as an addendum to his book, Arc Flash Calculation Studies. This was based upon 2 technical papers which were "Arc Flash Calculations for Exposures to DC Systems" by D.R. Doan and "DC Arc Models and Incident Energy Calculations" by R.F. Ammerman, T. Gammon, P.K. Sen and J.P. Nelson. Following on from Jim's lead, these methods were introduced by commercial software companies. The calculator is based upon this method.

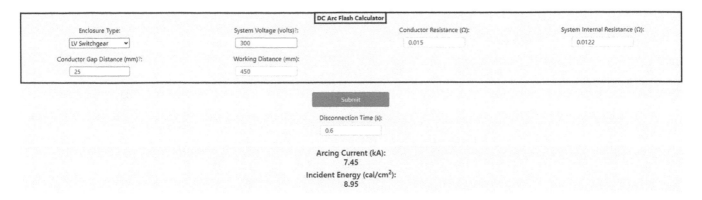

Figure 14.16 DC IE Calculator

Calculation method

If we take a simple DC circuit which is subject to an arc fault, the current that flows in the arc is found by application of ohms law which is $I = V/R$. So, the source voltage is divided by the sum of the series impedances made up of the; source resistance R_s, system (or conductor) resistance R_c and the arc resistance R_{arc}.

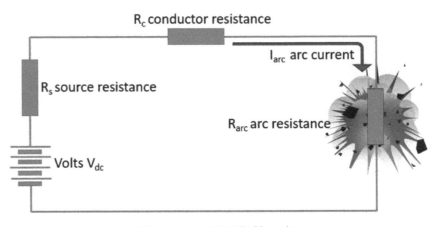

Figure 14.17 DC Circuit

To find the arc resistance R_{arc} we apply the following equation.

$$R_{arc} = \frac{[20 + (0.534 \times G)]}{I_{dc\ arc}^{0.88}}$$

Where G = conductor gap in mm and I_{arc} = DC arcing current

As can be seen we have two unknowns, R_{arc} and I_{dc_arc}. It is necessary to apply an estimate of arcing current of 50% of the prospective short circuit current in order to get an estimate of arcing resistance R_{arc}. This is the voltage divided by the sum of R_c and R_s which gives us the prospective short circuit current multiplied by 50% giving the estimate of arcing current. This estimate is entered into the above equation to give a first arc resistance. This is added to the actual values of R_c and R_s which reveals a total estimated impedance which is divided into the voltage once more to determine a new value of arcing current. This new arcing current is used in the above equation and once again a new value of arc resistance is determined. This is repeated and through a process of iteration steps, the calculated values of both arcing current and arc resistance will get closer to the previous values until there is no significant difference.

The next stage in the calculation process is to find the arc energy E_{arc} using the formulae $P = I^2 \times R \times t$ watt/seconds using the calculated values for arcing current I_{dc_arc} and arc resistance R_{arc}. The value t is the disconnection time of the circuit which is found by consulting the upstream protective device time current curve.

If the arc is in open air the incident energy is found by: $E_{i\ air} = \dfrac{E_{arc} \times 23.9}{4\pi \times d^2}$ cal/cm² where d: is the distance measured from the arc source to the worker.

If the arc is enclosed, then factors are selected in accordance with the following table from the Wilkins paper.

Optimal Values of a and k		
Enclosure Type	a (mm)	k
LV Switchgear	400	0.312
MV Switchgear	950	0.416
Panelboard	100	0.127

TABLE 14.18

The enclosed (box) incident energy is found by $E_{i\ box} = \dfrac{E_{arc} \times k \times 23.9}{a^2 \times d^2}$ cal/cm².

Figure 14.16 shows the dashboard from the DC incident energy calculator. An enclosure type or open air is selected form the drop-down field and this will require values for the internal resistance of the source, the conductor resistance and the arc gap. The Arcing Current $I_{dc\ arc}$ is then shown which can be used to determine the disconnection time from the TCC and entered into the disconnection time t. If this is not available or there is not a device in circuit, a value can be entered which is an estimate of reaction time. (note, the IEEE 1584 Guide uses a figure of two seconds if there is no restriction of access). Finally, a working distance d is entered which is measured to the source of the arc.

The calculated incident energy, which is shown, depends upon the type of enclosure (or air) distance and disconnection time given the electrical system parameters.

The entry and output field are listed as follows:

Enclosure Type

Choose from a drop-down menu as described in Table 14.18 previously.

System Voltage (volts)

Enter DC open circuit volts.

Conductor Resistance (Ω)

Enter the total conductor resistance from the source to the equipment in ohms.

System Internal Resistance (Ω)

Enter the internal resistance of the system or source in ohms.

Conductor Gap Distance (mm)

Enter the gap between conductors in mm.

Working Distance (mm)

The distance between the potential arc source (not just the front of the panel) and the face and chest of the worker.

DC Arcing Current I_{arc} (kA)

The calculated dc arcing current.

Disconnection Time (seconds)

This is the time to clear the dc arcing current I_{arc} and is obtained from the time current curve of the upstream protective device.

Incident Energy (cal/cm^2)

The calculated dc incident energy at the working distance.

References

1. Complete Guide to Arc Flash Hazard Calculation Studies by Jim Phillips P.E – Brainfiller Inc. 2010
2. DC Arc Flash Calculations by Jim Phillips – Posted on Brainfiller.com, November 20, 2019
3. Arc Flash Calculations for Exposures to DC Systems by D.R. Doan – IEEE Transactions on Industry Applications, Vol.46 No 5
4. DC Arc Models and Incident Energy Calculations by R.F. Ammerman, T. Gammon, P.K. Sen and J.P. Nelson – IEEE Transactions on Industry Applications, Vol.46 No 6
5. Electric Arcs in open Air by A.D.Stokes and W.T.Oppenlander – Journal of Physics D: Applied Physics 1991
6. Simple Improved Equations for Arc Flash Analysis by R. Wilkins – IEEE Electrical Safety Forum, August 30, 2004.

14.6 Blast Pressure Calculator (LEVEL 2)

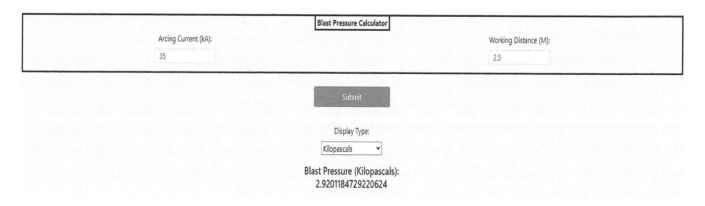

Figure 14.19 Blast pressure calculator

The arc blast pressure calculator theoretical formulae developed by Ralph H Lee, published in his paper "Pressures Developed by Arcs" in 1987. There is much debate about how Lee's theoretical formulae can be interpreted in the real world, but this is still widely cited by engineers when calculating the severity of the blast hazard. More information is given in Chapter 2: What is Arc Flash. This calculator uses his formulae to give an initial blast pressure figure. As can be seen in the screen shot of the calculator, the following details the data entry values that will be required.

Arcing current (kA)

Enter the calculated arcing current in kA. (Not the prospective short circuit current)

Working distance (metres)

Enter the working distance which is the distance from the prospective arcing source to the person's torso.

Blast Pressure

The initial blast pressure is given in Kilopascals, lbs/ft², n/m² and Atmospheres all of which can be selected via the drop-down list.

14.7 DGUV Box Test Algorithm Calculator (LEVEL 1)

This calculator in based upon the DGUV-I-203-5188 E Guide to the Selection of Personal Protective Equipment for Electrical Work. It is derived from the standard IEC 61482-1-2: Live Working - Protective Clothing Against the Thermal Hazards of an Electric Arc Method 2: Determination of arc protection class of material and clothing by using a constrained and directed arc (box test). Whilst the DGUV guide is used in parts of Europe for low voltage utility applications, I would recommend the use of IEEE 1584 for industrial and commercial purposes. There is more information about this standard and the DGUV Box Test Algorithm in Chapter 7: Protection which gives a detailed comparison to IEEE 1584 and Chapter 16: Rules, Codes and Legislation.

DGUV Calculator Notes

This should be read and applied in accordance with the German publication DGUV -I- 203-5188 E Guide to the Selection of Personal Protective Equipment for Electrical Work.

Requirements for use

Instead of creating an incident energy level in cal/cm^2 the calculator will produce an arc power based upon the physics of the arcing current.

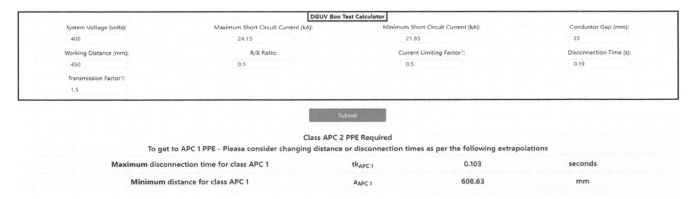

Figure 14.20 DGUV Box Test calculator

Voltage - DATA ENTRY

The three-phase line to line voltage in volts.

Maximum & Minimum Short Circuit Current (I''_{k3max} & I''_{k3min}) in kA - DATA ENTRY

A maximum and a minimum short circuit current is required in accordance with German standard VDE 0102:2016 short-circuit currents in three-phase AC systems. This is required in order to calculate a minimum short circuit current in order to determine the minimum arcing current. The standard is based upon DIN EN 60909 Short-circuit currents in three-phase AC systems. The standard DGUV -I-203-078-5188e states that "calculation software is usually available for this process". This is to obtain maximum and minimum prospective three phase initial short-circuit AC currents I''_{k3max} and I''_{k3min}. However, for single circuits at low voltage, the minimum and maximum short circuit currents are given in this guide in the prospective short circuit current calculator.

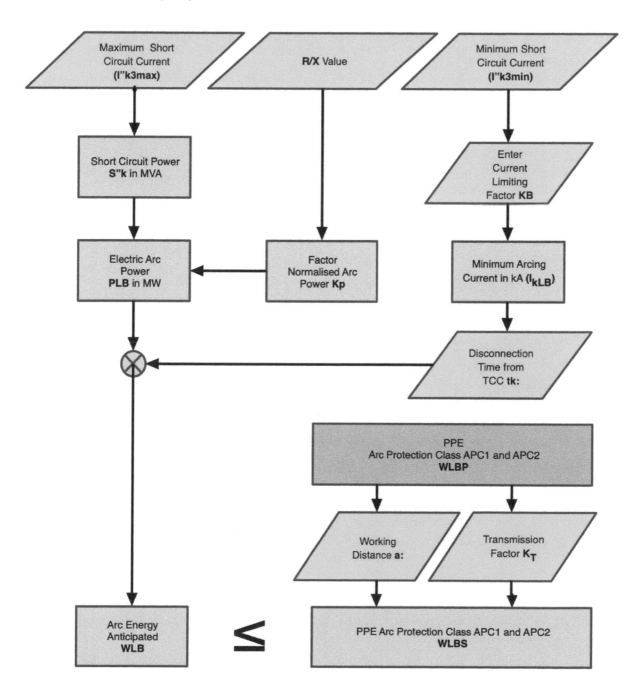

Figure 14.21 Flowchart

R/X ratio Data Entry

An R/X ratio is required which is not always easy to establish. However, the R/X can be found from the prospective short circuit current calculator as a reciprocal of the X/R value for low voltage circuits fed by a transformer.

Electrode Gap in mm (d) – DATA ENTRY

This is the minimum gap between the conductors measured in millimetres.

Current Limiting Factor (K_B)– DATA ENTRY

The current limiting factor is simply applied to the minimum short circuit current I"k3min to produce an arcing current.

Minimum Arcing Current in kA (I_{kLB}) – CALCULATED

This is calculated from the current limiting factor and minimum short circuit current.

Disconnection Time in seconds (tk)– DATA ENTRY

This is the time to clear the minimum arcing current above and is obtained from the time current curve of the upstream protective device. See Chapter 4: Prediction.

Short Circuit Power in MVA (S"k)– CALCULATED

This is simply the maximum available power based upon the maximum short circuit current I"$_{k3max}$ and line voltage. ($\sqrt{3}\ VI"k3max$)

Normalised Arc Power K_p - CALCULATED

The normalised electric power is a factor which accounts for the arc formation and geometry of the conductors. For worst case conditions it is calculated from the equation $kpmax = \frac{0.29}{(R/X)^{0.17}}$.

Electric Arc Power P_{LB} (MW) - CALCULATED

Electric arc power Is the power transferred to the arc based upon the Short Circuit Power (S"k) times the normalised electric power (K_p).

Arc Energy (Anticipated Value) W_{LB} - CALCULATED

The anticipated value of arc energy takes into account the electrical arc power and the disconnection time. $WLB = PLB \times tk$

Arc Protection Class of PPE W_{LBP}_APC1 & W_{LBP}_APC2- GIVEN VALUES

The arc protection class of PPE is given from the DGUV guide, and the latest figures are W_{LBP}_APC1 = 168 kJ and W_{LBP}_APC2 = 320 kJ. This value is the protection level of APC1 or 2 at a distance of 300mm. (The sample set up distance for the box testing of PPE)

Transmission Factor K_T - DATA ENTRY

The transmission factor is derived from whether the arc is enclosed (in a box), semi enclosed, in air but against a rear wall or in open air. A diagram of the three configurations is shown in Figure 14.22.

K_T Transmission Factor Reference Values

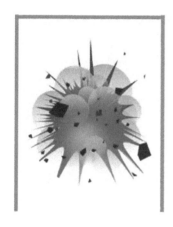

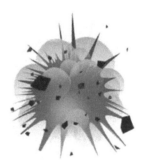

Box V=1.6l Side and Rear Walls Low Volume	Rear wall only or a large combustion volume	Very large combustion volume
K_T =1	K_T =1.5...1.9	K_T =2.4

Figure 14.22 Transmission Factors

Working Distance (a:) - DATA ENTRY

Distance in millimetres from the arc to the person's body (torso).

Protection Level of Clothing W_{LBS}_APC1 and W_{LBS}_APC1 - CALCULATED

Protection level of the clothing is obtained by extrapolation of the box test parameters to the arc location. $WLBS = Kt \times \left(\frac{a}{300}\right)^2 \times WLPB$

PPE Selection

This is the final step which is to compares arc energy (anticipated value) W_{LB} with the protection level of the clothing W_{LBS}. Protection is given if the following is true.

$$WLBS \geq WLB$$

14.8 Prospective Short Circuit Current Calculator Tool

Introduction

The prospective short circuit current calculator tool is useful to provide an estimate of the prospective short circuit current at the equipment where incident energy hazard evaluation is to be carried out. By using the calculator, the prospective short circuit current can be determined which can then be used in the Incident energy calculator.

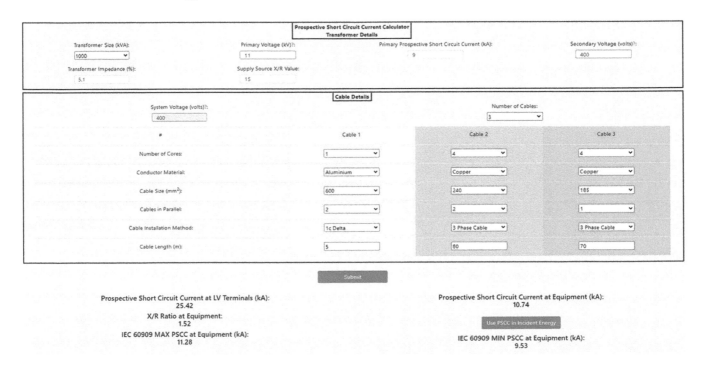

Figure 14.23 Prospective short circuit Current Calculator

The tool is based upon the impedance method which will provide an RMS symmetrical short circuit current which can be inputted directly into the IEEE 1584 Incident Energy Calculator. There is also a separate output of maximum and minimum fault currents in accordance with IEC 60909 Short-circuit currents in three-phase AC systems. This will be required for the DGUV-I 203-078 German Calculator if required.

The limitations are:

- Three-phase short circuits up to a maximum secondary voltage of 600 volts.
- Symmetrical three-phase fault current only.
- Single radial circuits only.
- Main's contributions only. Motor contribution can be included when using the IEEE 1584 Incident Energy Calculator.
- The cable resistance and reactance values are generic, users must satisfy themselves that these are a reasonable estimate.

The following diagram Figure 14.24 shows a single line representation of the components of the Prospective Short Circuit Current Calculator. In this case the equipment to be worked upon is represented by switchboard SB3 but this could alternatively be SB1 or SB2 simply by omitting the cable 3 or cables 2 and 3 respectively. The calculator should be used when equipment being studied has an upstream transformer device as the supply source or as an alternative, the transformer may be omitted to use the cable calculators only.

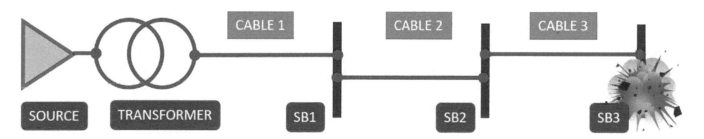

Figure 14.24 Single line diagram

Transformer

The transformer short circuit calculator, shown below, can be used to determine prospective short circuit current and X/R ratio at the low voltage terminals of a transformer. The generic transformer X/R ratios are taken from several sources of information including different manufacturers and also IEEE 242 which suggests typical levels. The author has taken these values to create a table of reasonable averages for distribution type power transformers based upon voltages up to 20kV primary and up to 600 volts secondary. Transformers will all possess different specifications due to the nature of their manufacturing process, so a higher degree of accuracy is possible by obtaining the original transformer specification. The calculator is, however, much more accurate than taking manufacturer's estimated LV fault data based upon infinite HV impedance and average % impedance values.

To use this calculator:

i. Inspect the transformer to be assessed taking note of kVA size, primary voltage, % impedance and secondary voltage. See Chapter 8: Data Collection for further guidance.

ii. Enter this information into the calculator by selecting/inputting the relevant data into the data entry fields.

iii. Contact the local utility supplier to determine the actual prospective short circuit current of the supply. It is always good practice to obtain the minimum fault level also as this could lead to a potentially longer disconnection time in the case of an arc flash.

As indicated in step 1 above the transformer impedance can be obtained from the transformer nameplate. Failing that, the following table 14.26 shows some standard impedance levels that can be used with a caution to apply some sensitivity validation (or what if analysis) to any incident energy levels obtained as a result.

kVA Rating of the MV/LV transformer	<630	800	1,000	1,250	1,600	2,000
Impedance volts %	4	4.5	5	5.5	6	7

Figure 14.26 Standardised short circuit voltage for public
distribution transformers - Source Schneider Electric.

The calculator uses the following table, Figure 14.27 of average X/R ratios for transformer kVA sizes as described previously.

The calculator will then produce results for the prospective short circuit current and the X/R ratio at the Low Voltage terminals. These values can then be used in the fault current level calculator to make allowances for cabling between the transformer and the equipment.

Please note that this tool is valid only for transformers up to 20kV on the primary side and up to 600V on the secondary side only. For transformers above either of these values, contact the manufacturer for X/R values. Note also that this calculator assumes a typical high voltage source X/R value of 15. Although this is a fair assumption for most situations, this still would require consideration to ensure that the calculator's assumptions are correct for the specific transformer being assessed.

This calculator should be used when equipment being studied has an upstream transformer or low voltage mains from a utility company as the supply source. By using the calculator, the prospective short circuit current can be determined which can then be used in the incident energy calculator.

kVA	X/R
112.5	2.5
150	3.0
225	3.25
300	3.5
400	3.6
500	3.7
630	4.0
750	4.6
800	4.7
1000	5.1
1250	5.4
1500	5.6
1600	5.7
2000	6.0
2500	6.3
3000	6.6

Table 14.27 X/R Ratios

Cables

The calculator tool shown is useful to provide an estimate of the prospective short circuit current at the equipment based upon Schneider Electric Cahier technique no. 158 "Calculation of short-circuit currents" by Metz-Noblat, Dumas and Poulain. This is a very useful technique to provide conductor resistance and reactance values based upon material resistivity calculations and average reactance values. When calculated to this method, the results are very similar to the generic cable resistance for non-flexible cables (stranding class 2) found in BS EN 60228 standard on conductors of insulated cables. The tool requires information about the system voltage, source prospective short circuit current, X/R ratio and details about the cables between the supply and the equipment itself. The prospective short circuit current is based upon the cable temperature at 20°C.

The prospective short circuit current at the equipment can be found by:

i.   Determine the system voltage, source prospective short circuit current and the X/R ratio and enter the data into the calculator. This is not required if the transformer calculator has been used above.

ii. Assess the cabling between the source and the equipment noting conductor material, number of cables in parallel, number of cores for each cable, cable size and length. Enter the data into cable 1 by selecting from the drop-down lists provided. See Chapter 8: Data Collection for further guidance.

iii. If there are multiple cable types in series from the source to the equipment, take note of each cable's parameters as in step 2 and enter the extra cable data into the optional cables 2 and 3 by selecting the relevant drop-down list options provided.

After the above steps have been followed, the tool will generate an estimated prospective short circuit current at the equipment which can be used in the incident energy calculator or other estimation modules provided.

The cable resistance and reactance values are generic and whilst this will allow for a reasonable estimate, cable manufacturing processes and materials will inevitably lead to variances. The generic values are given below in Figure 14.28 and users are advised to satisfy themselves of their accuracy.

Cable Installation Method	Composite 3 phase Cable	Trefoil or Delta Single Core Cables	Single Core Cables Touching	Single Core Cables Trunking	Single Core Cables Spaced 2X or 4X Radius
Diagram	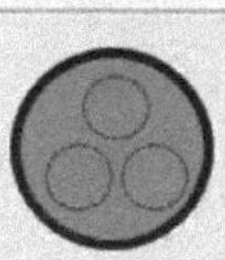		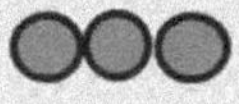	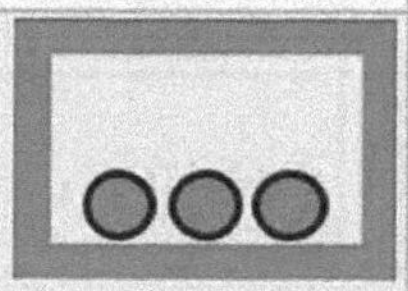	
Generic Average Reactance (milliohms/metre)	0.08	0.085	0.095	0.135	0.145 2X Radius 0.190 4X Radius

Figure 14.28 Average Reactance Values - Source Schneider Electric.

As can be seen, the reactance values can vary widely with different installation methods and any electrical engineer who is carrying out retrospective system studies needs to be aware of this fact.

The resistance values are based up $\rho 0$ = resistivity of conductors at 20⁰C = 0.01851 Ω mm²/m for copper and 0.02941 Ω mm²/m for aluminium. For loaded conductors in order to obtain a minimum short circuit current in accordance with IEC 60909, see later paragraph, a multiplication factor of 1.25 has been used. This factor will return a value which is in line with a conductor operating temperature maximum of 90⁰C which should be conservative for power systems.

Please note that this tool is only valid for system voltages between 208 – 600V (3 phase AC) and a source prospective short circuit current of between 500 A to 106 000 A.

IEC 60909 Maximum and Minimum Prospective Short Circuit Current

A further feature of the calculator is that you can obtain a maximum and minimum prospective short circuit current in accordance with IEC 60909: Short circuit currents in three-phase AC systems. To use the DGUV Box Test Algorithm it will be necessary to have both maximum and minimum figures and the voltage correction factors given below are as stated in DGUV I 5188E. The IEEE 1584 guide calculates an arc variation factor which will reduce the arcing current in order to take into account fluctuations in fault currents.

This is calculated as follows.

- The maximum short circuit current is found by applying a voltage factor c to the estimated prospective short circuit current above which was based upon conductor temperature of 20⁰C. The voltage factor is based upon a 6% tolerance giving a factor of c_{max} =1.05.
- The minimum short circuit current is found by calculating the estimated prospective short circuit current based upon fully loaded conductors which are operating at design temperature maximum of 90⁰C and the applying a voltage factor c_{min}. The voltage factor is based upon a 6% tolerance giving a factor of c_{min} = 0.95.

To summarise, the following is a list of data entry and output fields from the calculator.

TRANSFORMER DETAILS SECTION

Transformer Size (kVA)

Select the transformer size from a dropdown list of distribution transformers from 112.5 kVA up to 3000 kVA. If there is no transformer and the source is at low voltage from a utility company, select "NA" and go directly to the cable details section.

Primary Voltage (kV)

Enter the primary voltage to the transformer up to a maximum of 20 kA.

Prospective Short Circuit Current (kA)

Enter a figure for the prospective short circuit current at the primary of the transformer.

Secondary Voltage (volts)

Enter the secondary three-phase line to line voltage at the transformer low voltage terminals between 208 and 600 volts.

Transformer Impedance (%)

Enter the % transformer impedance figure which can be found on the transformer nameplate.

Supply Source X/R Value

The calculator assumes a typical high voltage source X/R value of 15 but a specific X/R ratio can be entered.

CABLE DETAILS SECTION

System Voltage (volts) (No transformer)

If there is no transformer used, enter the low voltage three-phase line to line system voltage in volts.

Enter Source Prospective Short Circuit Current (kA) (No transformer)

If there is no transformer used, enter a figure for the prospective short circuit current for the low voltage source.

Supply Source X/R Value (No transformer)

If there is no transformer used, enter a source X/R value, otherwise a default value of X/R = 1 will be used.

Number of Cables

Enter the number of cables that will be used between the supply source and the equipment up to a maximum of three. A data entry block will be displayed for each cable. Cable one will always be the nearest to the supply source as in the single line schematic shown earlier.

Number of Cores

Enter the number of cores for each cable used.

Conductor Material

Select conductor material as aluminium or copper from the drop-down field.

Cable Size (mm²)

Choose a cross sectional area of each conductor from the drop-down list from between 16 mm² up to 1000 mm².

Cables in Parallel

Choose the number of cables installed in parallel from the drop-down list.

Cable Installation Method

Choose the cable installation method from the drop-down list.

Cable Length (m)

Enter the length of the cable in metres.

OUTPUTS

Prospective Short Circuit Current at LV Terminals (kA)

This is the symmetrical prospective short circuit current at the low voltage source, transformer low voltage terminals or low voltage utility mains.

X/R ratio at Equipment

This is the X/R ratio at the Equipment.

Prospective Short Circuit Current (kA)

This is the symmetrical prospective short circuit current at the equipment after calculating impedances of the supply source and all cables. This figure can be used to input directly into the IEEE 1584 incident energy calculator by selecting the "Use PFC in Incident Energy" option button.

IEC 60909 MAX PSCC at Equipment (kA)

As described in this chapter an output is given for the symmetrical maximum prospective short circuit current as calculated using IEC 60909 correction factors.

IEC 60909 MIN PSCC at Equipment (kA)

As described in this chapter an output is given for the symmetrical minimum prospective short circuit current as calculated using IEC 60909 correction factors.

14.9 LV HRC Fuse Time Current Curve Tool

Introduction

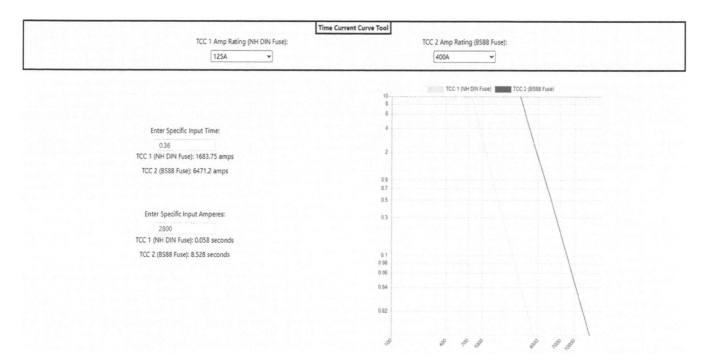

Figure 14.29 LV HRC TCC Calculator

The time current tool provides generic clearance curves for the DIN NH and BS88 industrial fuses to the standard IEC 60269. The calculator produces a time current curve characteristic plot for up to two curves. The data entry and output is detailed as follows.

TCC 1 Amp Rating (NH DIN Fuse)

A drop-down list menu gives a range of choices between 32 amperes and 1250 amperes for NH DIN Fuses.

TCC 2 Amp Rating (BS 88 fuse)

A drop-down list menu gives a range of choices between 32 amperes and 1250 amperes for BS 88 Fuses.

TCC Chart

This will display both TCC 1 (NH DIN Fuse) and TCC 2 (BS88 Fuse) fuse time current curves. Either one can be deselected by clicking on the associated colour block at the top of the graph.

Enter Specific Input Time

By entering a specific input time ranging from 0.01 to 10 seconds), a corresponding clearance current for the associated fuse will be given. (This value will be displayed in amperes)

Enter Specific Input Current

By entering a specific input current in amperes (guidance ranges will be displayed depending on the selected fuse amps), a corresponding clearance time for the associated fuse will be given.

Notes:

A mistake that is often seen is the use of pre-arcing curves with no allowance for tolerance or clearance. This calculator factors this into the curves and is based upon generic curves as displayed in Amtech for BS 88 fuses. The DIN fuses are based upon the ABB DIN-type HRC-fuse links, <1250 ampere gG -types from the 2018 Protection & Connection Fuse Gear catalogue. Allowance for clearance has been take from the ABB "Curves" software.

14.10 IDMT Time Current Curve Calculator

The generic Inverse Definite Minimum Time (IDMT) time current curve calculator will allow you to not only produce curves for standard IEC and IEEE relay characteristics but will give a trip time for a given arcing current.

The standard IDMT overcurrent characteristics are expressed as a mathematical formula according to the IEC 60255-3/BS142 standards and IEEE C37.112-1996 and are often the basis of the electromechanical time current curves. The fact that the curves are based on a mathematical formula makes it easy to sense check the relay settings and of course electronic relays use the same curves.

The equation for operating time from IEC 60255 trip curves is as follows.

$$t(I) = TMS \left\{ \frac{k}{(\frac{I}{Is})^\alpha - 1} \right\}$$

Where, I_s = the current setting and I = the actual current and k and α are the curve type constants. See table below.

There are four curve types used in IEC 60255 which are: standard inverse, very inverse, extremely inverse and long time standard inverse. For electromechanical relays, the curves are fixed for the particular relay model. So, in other words, if a standard inverse function is required then that is built into the specification of the relay. It will be necessary therefore to check the relay model with manufacturer's data to determine the curve type.

CURVE TYPE	Constant	
	k	α
Standard Inverse	0.14	0.02
Very Inverse	13.5	1
Extremely Inverse	80	2
Long Time Standard Inverse	120	1

TABLE 14.30 IEC CURVE CONSTANTS

Note that US relay characteristics to ANSI and IEEE follow a similar format to the above equation but have a further constant B which is added to the expression.

The equation for operating time from IEEE C37.112-1996 trip curves is as follows.

$$t(I) = TD \left\{ \frac{A}{\left(\frac{I}{Is}\right)^p - 1} + B \right\}$$

Where, I_s = the current setting and I = the actual current and A, B and p are the curve type constants. TD stands for time dial and is often referred to as a time dial multiplier or TDM. The relationship is: . See table below.

CURVE TYPE	Constant		
	A	B	p
IEEE Moderately Inverse	0.0515	0.114	0.02
IEEE Very Inverse	19.61	0.491	2
IEEE Extremely Inverse	28.2	0.1217	2

TABLE 14.31 IEEE CURVE CONSTANTS

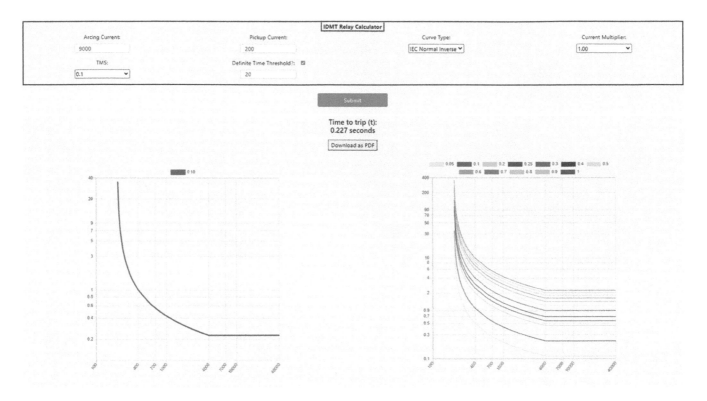

Figure 14.32 IDMT Time Current Calculator

The following is a description of the user entry and calculation field given in the image of the IDMT time current curve calculator given in Figure 14.32.

Pickup current. (I*s*) DATA ENTRY

This is the value at which the relay will start to operate and is usually the current transformer primary turns ratio. So, if this is 400/5 for instance, enter 400. This does of course assume that the relay matches the secondary winding of the current transformer.

Curve type

Choose from the drop-down menu from the following options. IEC standard inverse, IEC very inverse, IEC extremely inverse, IEC long time standard inverse IEEE moderately inverse, IEEE very inverse and IEEE extremely inverse.

Current Multiplier. (x I*s*) (Range: from 0.5 to 2 in 0.05 intervals)

Enter a value that corresponds with the relay setting. For instance, if the relay is a static, digital, or numerical type of relay this can be obtained directly from the relay setting sometimes via slider switches or digital readout. So, a current multiplier of 1 will be the pickup current x 1 or 100%. For electro-magnetic relays the detail can be obtained from the plug bridge and the tapping range is usually 50%, 75%, 100% (For a 5-ampere relay, 100% means 5A), 125%, 150%, 175% and 200%. Some manufacturers may use 1.5, 3.5, 5, 6.25, 7.5, 8.75 and 10A. For example, for a CT rating of 100/5A, if the relay is set to operate at 5A then the plug setting will be equal to relay current setting/5A =

5A/5A = 1 or 100%. For a relay to operate at 2.5A, the plug setting (for this example) will be 2.5A/5A = 0.5 or 50%.

TMS (Time Multiplier Setting)

Choose from the drop-down list from between 0.05 and 1. For static, digital, or numerical type of relay this can be obtained directly from the relay setting sometimes via slider switches or digital readout. For electro-magnetic relays, the detail can be obtained from the time dial on the front of the relay, this is shown in Chapter 8: Data Collection section 8.4.2 Electromechanical Inverse Definite Minimum Time Relays.

Definite Minimum Time

This field gives the ability to apply a definite minimum time for operation of the relay when the tick box is checked. Different relays give a different threshold for minimum time for operation which, when stated, are usually at 10 to 40 times the pickup current as adjusted by the current multiplier. So, for instance, if the pickup current is 100 amperes, the current multiplier is 2 and the definite time threshold is 20, then the curve will follow the relay curve characteristic up to 4000 amperes (100 x 2 x 20). Above that threshold, the time to trip will remain static or in other words a definite minimum time for operation. There will be no definite time for operation if the tick box is left unchecked.

The effect of the current multiplier, the time multiplier setting, and definite time threshold can be shown in Figure 14.33 below. So, in this example, the pickup current is 100 amperes, the current multiplier is 2 and the definite time threshold is 20. So, the set pickup is 200 amperes (100 x 2) and the definite time threshold is 4000 amperes (200 x 20).

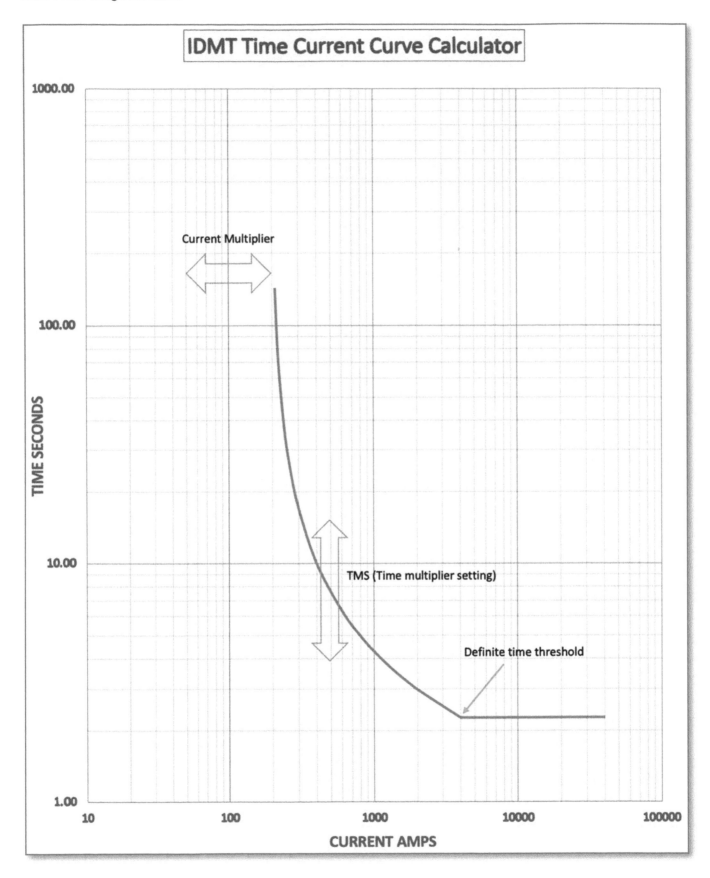

Figure 14.33 IDMT Effect of settings.

Arcing Current

Enter the arcing current from the IEEE 1584 incident energy calculator (calculator 1) both for actual and reduced values. Clearly, the arcing current must be greater than the pickup current as adjusted by the current multiplier for the relay to operate.

Outputs

A time current curve for the full range of time multiplier settings will be displayed for the relay data entered when the submit button is selected. Curves can be deselected by clicking on the associated colour block at the top of the graph. In addition, a time to trip based upon the arcing current given will be displayed. The arcing current can be changed to consider the minimum arcing current from the IEEE 1584 incident energy calculator to take account of possible increased trip times.

Allowance for relay error and switchgear operating time

Do not forget the relay error and switchgear operating time. Trip times are made up of relay reaction time + trip coil delay time + circuit breaker mechanism operating time. All relays have errors in timing which have to be declared if manufactured and tested in accordance with IEC 60255. We must also factor in the time for operation of the switchgear. This again can be obtained from manufacturer's data, commissioning, or timing tests but as a guide for the purposes of an arc flash calculation a generic allowance can be built in. A table of allowances and generic switchgear operating times are given in Chapter 8: Data Collection.

14.11 Risk Assessment Forms

Within the online guide (https://www.ea-guide.com), the results of the calculations can be printed off. There is an option to create dynamic risk assessment forms in line with the 4 P principles. The following is an example of one such risk assessment form which is provided for the IEEE 1584 calculator.

EAG European Arc Guide
Simple & Accessible
Arc Flash Risk Assessment

Client

Task

DATE

Predict Section

Voltage
Prospective Short Circuit Current
Working Distance
Electrode Configuration
Electrode Gap
Enclosure Size
Motor Contribution Details
Protective Device Details   Disconnection Time
Maximum Interrupting Rating
Predicted Incident Energy at above Working Distance
Arc Flash Boundary

Prevent Section

Work to be carried out DEAD (Note that the process of de-energisation may include increased arc flash risks or potential exposure). **MUST BE IMPLEMENTED WHENEVER POSSIBLE**	YES	NO
Upstream circuit breaker settings may be changed to reduce the arc flash incident energy	YES	NO
Upstream fuse may be changed to reduce the arc flash incident energy	YES	NO
The upstream protective device has been adequately maintained to my satisfaction	YES	NO
Switching sequences have been correctly identified to minimise risk and incident energy levels	YES	NO
Remote switching has been considered	YES	NO

Detail here any other measures taken to reduce or minimise incident energy levels:

Process Section

The workers are competent for the task	YES	NO
The working space is adequate to allow workers to step back without danger	YES	NO
If the work is to be carried out at height, there are adequate measures available to reduce the risk of falling	YES	NO
Accompaniment is required	YES	NO
The area is free from ignition hazards	YES	NO
There are no slipping or tripping hazards in the working zone	YES	NO
There is adequate lighting at the point of work	YES	NO
The equipment is shrouded to EN60529 ingress protection IP2X, or greater	YES	NO
If not, can temporary insulation or shrouding be applied safely to provide additional protection?	YES	NO
The equipment has been examined for damage or arcing using Senses; such as hearing, sight, smell?	YES	NO
There are no signs of burning / arcing / discolouration	YES	NO
There are no signs of moisture or dust ingress	YES	NO
There is no possibility of vermin or bird ingress	YES	NO
Tools to be used are insulated and suitable for the task	YES	NO
Instruments to be used have been inspected for damage and correct specification for transient over-voltage levels	YES	NO
Test leads are fused and in good condition	YES	NO

List here the specific actions required or undertaken as identified above:

Protect Section

The following measures have been selected from the Protect section in order to provide protection for the worker against the residual risk.

The hazards and risks associated with the arc flash hazard for the specific task have been examined, adequate control measures put in place, and the work is safe to proceed as planned.

Approved

Date

Figure 14.34 Risk Assessment Form

Chapter 15

Why Do Arc Flash Calculations?

Introduction

I felt compelled to write this chapter after listening to a colleague who questioned the need for carrying out arc flash calculations. He mistakenly thought that calculations = PPE which cannot be further from the truth. Here are ten solid reasons for carrying out calculations in order to prevent the hazard from causing harm.

15.1 Root out danger

Arc flash is just another hazard and I have been consistent over the years that the major consideration is to understand it and root out real danger. The analogy of the hidden raging tiger lurking inside the electrical equipment works well for me as my prime consideration is the identification and evaluation of the hazard as the first steps in risk assessment. Calculation forces us to have a look at the upstream protective devices and in the same way as protecting against electrical shock, impedance is the enemy of good protection. So, we need to consider whether an arcing fault is going to be cleared quickly with minimal damage or, the fault will persist with catastrophic harm to the equipment, and anyone stood nearby.

15.2 Keeping people out of restrictive PPE

The difference between moon suits and no PPE at all is often just a few amps. In most of the studies that I have carried out, I can point out where 90% of high hazard levels could be averted by a competent engineer reconfiguring circuits or in most cases protection alterations. In one such case I was working for one of the largest technology factories in the UK where there were extremely high specification clean room environments. Following the initial calculations, it was discovered that there were 120 low voltage panels which were identified in the clean room production areas, that had high levels of incident energy. We managed to successfully reduce the levels at all these panels to below 1.2 cal/cm^2 at a distance of 450mm. Measures such as new protection settings, replacement of circuit breakers, installation of additional local switch fuses and re-feeding equipment were all adopted with relatively little cost and effort. The motivation there was partly reflected in the fact that clean room flame resistant PPE is very expensive but why should this approach not be applied to all industrial settings? This was carried out in 2008 and for such a large site, their safety performance has since been excellent.

15.3 The first step in risk assessment

When carrying out a risk assessment, as a minimum we must:

1. **Identify what could cause injury (hazards)**
   - This is derived from system parameters such as voltage, fault level and protection arrangements.
2. **Decide how likely it is that someone could be harmed and how seriously (the risk)**
   - This is derived from system conditions such as the condition of the equipment, the quality of the installation, how well it has been maintained and whether it is being operated in accordance with its original design and:
   - It is also directly related to the task to be performed.
3. **Take action to eliminate the hazard, or if this is not possible, control the risk.**
   - se the 4P approach to eliminate or control the risk.

Where we have the means to evaluate the hazards in the workplace, we must take advantage. As I have said earlier, it is a matter of a few amperes that can be the difference between an arcing fault being cleared quickly and harmlessly to a situation that results in catastrophic damage, losses and severe personal injury.

15.4 Examples of where calculations would have prevented accidents

I give several examples of failure of protection systems in Chapter 12: Myths and Mistakes where calculations would have picked them up in my view. There is the catastrophic failure of a capacitor that raised a main substation to the ground and caused an environmental incident and also the apprentice who was burned whilst connecting cables into a distribution board. Not featured in Chapter 12: Myths and Mistakes is an 11kV circuit breaker which exploded at a waterworks with no injuries but the costs were huge. All three incidents would have been averted had the upstream protection arrangements been examined as part of an incident energy study.

15.5 Helps to comply with the law

Whether it is the European Framework Council Directive 89/391/EEC (Workplace Health and Safety Directive) or the UK Management of Health and Safety at Work Regulations 1989 or the Irish Safety, Health and Welfare at Work (General Application) Regulations 2007, the cornerstone is risk assessment which means that any employer must evaluate all the risks to the safety and health of workers. Arc flash risk assessment for workers who operate in proximity to or on energized electrical equipment, cables and overhead lines, is an essential part of electrical safety management. Calculations give us the means to evaluate the hazard and also the means to provide prevention techniques such as automatic disconnection of supplies for arcing faults.

15.6 90% of the effort for a study is what we should be doing anyway

When carrying out system studies, the vast majority of the basic electrical system information gained is what any self-respecting electrical engineer ought to have at his or her fingertips. A precursor to an arc flash study is to have a single line diagram of the electrical system, a short circuit study and a protection coordination study. Appendix A from IEEE 1584 2002 described all the data that would be required in order to undertake an arc flash study. It broke the data down into four separate categories: 1. The diagram, 2. Short circuit study, 3. Coordination study and 4. The arc flash study. I have represented the data headings into a chart shown in Chapter 12: Myths and Mistakes and as demonstrated, the amount of effort required for the arc flash study is less than 10% of the total. That means that the vast majority of information is what most engineers who have responsibilities for an electrical installation will choose to keep. I would certainly feel uncomfortable as a duty holder if I did not have these records in place.

15.7 Design problems are averted

I describe in Chapter 12: Myths and Mistakes a fault level study, protection coordination study and an arc flash study that I carried out for a large teaching hospital about 10 years ago. I uncovered a design flaw where 2 MVA voltage optimisation units had been installed into the busbars between the two main transformers and the main LV switchboards for the critical care block. The additional dynamic impedance meant that in the event of an arcing fault, protection arrangements would not operate, and also downstream protection could be compromised as a result. The results were frankly quite startling, and I calculated that the incident energy at the low voltage switchboards could be as much as 73 cal/cm2 against 5.1 cal/cm2 for not having the units in service. Because of this and also the vulnerability of the patients in this unit, the units were removed at great expense. To have carried out the study at design stage, would have revealed this but since only zero impedance faults are considered in present building services standards, an opportunity was missed.

15.8 Providing knowledge to workers

I believe that information and training can make a step change in safety practices to reduce harm from arcing faults. Electrical workers are often the ones who suffer from arcing incidents and yet they know surprisingly little about the phenomena. For instance, the mistaken belief that low fault level means less harm can lead to a false sense of security. There may be others out there who remember the GEC videos from over 50 years ago showing switchgear explosions and how BS 88 fuses reduced the damage to zero by disconnecting high fault currents very rapidly. The video shows the catastrophic damage from impressive explosions if the fuse links were replaced with solid copper links at a prospective short circuit current level of 38,000 and 60,000 amperes. The fuse links disconnect the fault current in less than one half cycle. The reason for this is that the fuse operates in its current limiting zone. The diagram in Figure 15.1 shows that for arcing faults, very little incident energy will be generated IF there is enough prospective short circuit current available to rupture the fuse quickly. But, as can be seen, fault levels below 12kA will result in high incident energy which can cause serious damage and harm to workers nearby.

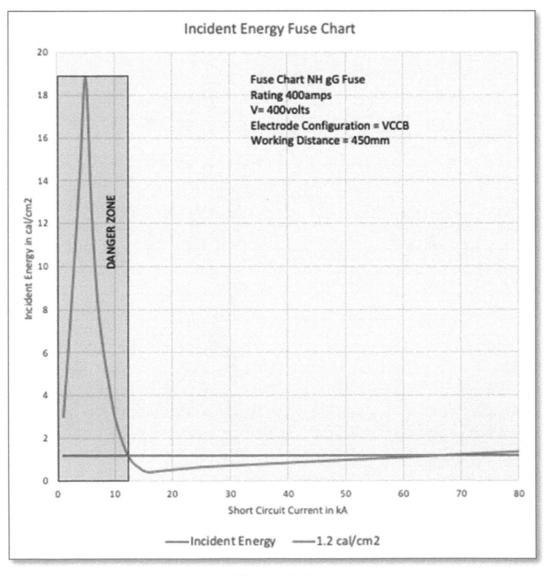

Figure 15.1

As is the case with shock protection, circuit impedance can be the enemy to safe automatic disconnection of supply. The electrician or engineer knows the voltage inside that control panel or switchgear so why should he or she not know the arc flash hazard level. By the way, the GEC video is still available on YouTube. Just enter "BS88 HRC fuse tests" into your search engine.

15.9 An essential tool for the designer

The designer does not necessarily need to know the actual incident energy or arc flash boundary but definitely needs to have confidence that his or her electrical protection arrangements will clear dangerous faults. This includes arcing faults! For traditional overcurrent protection, the information that will be required to fulfil this duty will be the amount of current that will flow in a fault. With that information, it can be assured that the protective device will operate in the instantaneous zone of the time current characteristic wherever that is possible. A few amperes can be the difference between safe clearance or destructive energy let through.

Predicting the severity of the arc hazard, including the arcing current, has been made more reliable in recent years through the publication of IEEE 1584 Guide for Performing Arc-Flash Hazard Calculations 2018. It is an auditable standard and widely accepted in the electrical engineering community. Two hundred technical experts were involved in the development of the guide over many years and was based on over 1860 tests performed at different voltage levels in high current laboratories. To quote from the scope of the Guide: "The purpose of the guide is to enable qualified person(s) to analyse power systems for the purpose of calculating the incident energy to which employees could be exposed during operations and maintenance work". It gives no recommendations for PPE. IEEE 1584 is therefore the flagship standard for determining the incident energy (thermal hazard) from an arc at three phase voltages in the range 208 volts to 15,000 volts AC.

(For the UK. Because designers now have the tools to predict arcing currents nowadays, I believe that there is less of a defence under EAWR Regulation 11 "Means for protecting from excess of current" as described in in the commentary in HSR 25.)

15.10 Improves electrical safety management

My strongly held belief is that by carrying out arc flash calculations, that it will inevitably force a fresh view of many of the components of a good electrical safety management program. In order to carry out the calculations effectively you will look at the equipment to be worked on or near which usually means control panels and switchgear in industrial settings. Purely on the basis of discovering system parameters you will need to know what the fault levels are and what is the strength and capability of the equipment and upstream protective devices. The big question that will crop up is, are you competent that the protective systems will operate in accordance with design parameters?

The likelihood of the hazard causing harm will be influenced by the conditions of the electrical system or equipment and also the task to be performed. Set this against the guidance of Chapters 5 and 6, Prevention and Process and it will become apparent that there is often a need to take immediate action to prevent unacceptable risk. Calculations will force you to look closely at equipment like process control panels which often highlights neglect which is a barometer of whether electrical safety policies are working or not. What I mean by this is when control panels show neglectful practices such as missing insulation, isolator control rods, trunking lids, dilapidation through water ingress and temporary lash up repairs, that will give an insight into the effectiveness of electrical safety policy.

All this will need to be fed back into policy and procedures in the form of a review which is what we should be doing on a regular basis. From that review, amendments to safety rules and training of staff will follow.

Learning Points

➢ There is more to arc flash risk management than PPE.
➢ Design Engineers should consider arcing faults on large power systems.
➢ Calculations of the arc flash can prevent accidents.
➢ It is possible to design out the need for heavy duty PPE.

Chapter 16

Rules, Codes & Legislation

Introduction

Arc flash is a global hazard and countries continue to develop safety practices to protect their workers. In Europe, arc flash requirements are much less prescriptive than in the United States and Canada. That is true when it comes to specific electrical codes of practice such as the National Fire Protection Association NFPA 70E "Standard for Electrical Safety in the Workplace" in the USA and CSA Z462, "Workplace Electrical Safety Standard" in Canada. Having consulted for companies that are US/Canadian owned, I have come across lots of situations where there may be head office directives to comply with NFPA 70E. Experience has shown that accommodations can be made to ensure compliance with local law and there is some very useful information in NFPA 70E if used with caution.

There is also a perception that there is a greater need for justifying work on or near live conductors in Europe than in the US. I would have agreed to this statement up until about 5 years ago. I co-presented a paper called "A European View of Arc Flash Hazards and Electrical Safety" to 500+ delegates at the IEEE Electrical Safety Workshop in Daytona Beach in Florida back in 2012, which effectively showcased a risk assessment approach based upon the legal requirements within the UK and Europe. After that conference I was asked to peer review a paper which approached the risk from ALARP principles. I am pleased to say that the US/Canadian approach has moved much more towards European risk assessment methodology since then.

Much of the law governing the arc flash hazard in Europe comes out of European directives with specific areas such as management of health and safety, use of work equipment, signs and signals and personal protective equipment directives. Within the framework of these directives, member countries pass their own legislation. Some legislation can be fairly unique such as Great Britain's Corporate Manslaughter and Corporate Homicide Act 2007, which went into effect on April 6, 2008. This clarifies the criminal liabilities of companies including large organizations where serious failures in the management of health and safety result in a fatality. The following gives an overview on the various European directives and standards and there is also information on how the directives are represented within Irish and UK law. Although the UK has left the EU, there are no plans to overhaul the present legislation that is based upon these directives, that the UK helped to draft, in respect of electrical safety. Should that change, this guide will be updated.

16.1 Europe - Electrical Safety

There is no doubt that firm prescriptive guidance for arc flash in Europe is not as advanced as in the United States where the National Fire Protection Association's NFPA 70E Standard for Electrical Safety in the Workplace is in existence. Naturally this does not mean that the problem of arc flash does not exist nor is it unacknowledged, it is more a case of an approach to hazard and risk that is different.

16.1.1 European Union

The European Union consists of 27 member countries with over 500 million citizens speaking 23 different official languages and providing for over a quarter of the global economy in terms of GDP. Laws that are passed as regulations or directives within the EU are mandatory across all member states and form the starting point for electrical safety compliance.

16.1.2 Framework Directive

In 1989, what would be known as the "Framework Directive" European Council Directive 89/391/EEC (Workplace Health and Safety Directive) was passed which introduced measures to encourage improvements to the safety and health of workers. A cornerstone of this directive is risk assessment which means that there are no prescriptive standards covering electrical safety, especially arc flash, and certainly no link from the hazard directly into arc flash PPE.

Directive 89/391/EEC creates an obligation on behalf of the employer to assess the level of risk involved in the workplace and the effectiveness of the precautions in place. For electrical work, all hazards should be considered, including the arc flash hazard and not purely shock, as is often the case with many European companies. Arc flash risk assessment for workers who operate in proximity to or on energized electrical equipment, cables and overhead lines, is an essential part of electrical safety management. Electrical work has to be carried out with conductors de-energized and isolated wherever possible and there is a very low tolerance for live working in many European countries.

As a result, many manufacturing plants restrict live working to inspection, diagnostic testing and commissioning purposes. There are, however, many tasks that require working either on, or in close proximity to, energized equipment. Even then it should also be acknowledged that the process of de-energization often requires exposure to the hazard through interactions such as switching, racking and testing of equipment.

16.2 Other EU Directives

The following is a list of EU directives that have most impact, in addition to the Framework Directive, on the management of the arc flash hazard.

16.2.1 PPE Directive

European Council Directive 89/656/EEC covers the minimum health and safety requirements for the use by workers of personal protective equipment at the workplace.

The PPE Directive sets the minimum requirements for the assessment, selection and correct use of personal protective equipment. Priority must be given to collective safety measures. PPE can only be used where the existing risks cannot be sufficiently limited by technical means or collective protection or work organization procedures. The employer must also provide the appropriate equipment free of charge and ensure that it is in good working order and hygienic condition.

PPE can only be prescribed after the employer has analysed the risks which cannot be avoided by other means. For arc flash this means an employer must consider other means of achieving safety prior to considering the use of PPE such as the elimination of hazard, engineering controls and safe systems of work. Issues of use, which must be part of the risk assessment process, must include ergonomics, sensory deprivation of user, continuing integrity of PPE and other injury mechanisms, including loss of hearing and sight. In summary, PPE must be used as a last resort.

16.2.2 Use of Work Equipment Directive 2009/104/EC

The Use of Work Equipment Directive 2009/104/EC defines the minimum health & safety requirements for the operation of work equipment by employees at work.

The employer is obliged to take every measure to ensure the safety of the work equipment made available to workers. During the selection of the work equipment the employer shall pay attention to the specific working conditions which exist at the workplace, especially in relation to the health and safety of workers. If risks cannot be fully eliminated during the operation of the work equipment, the employer shall take appropriate measures to minimize them. Employers have to provide workers with adequate, comprehensible information, usually in the form of written instructions, not only for the use of the equipment but also foreseeable abnormal situations. Workers shall be made aware of dangers relevant to them. The employer shall ensure that workers receive adequate training, including risks and specific training on specific risk equipment.

Arc flash risk enhancing situations such as the lack of proper equipment maintenance, inadequate equipment design and installation, deficient commissioning and initial inspection and poor worker competence should not happen if working within the "Use of Work Equipment Directive".

16.2.3 Provision of Health and Safety Signs at Work

Council Directive 92/58/EEC defines the minimum requirements for the provision of health and/ or safety signs at work and complements the Framework Directive 89/391/EEC on health and safety at work.

This directive states that safety and/or health signs must be provided where hazards cannot be adequately reduced by techniques for collective protection or by measures, methods or procedures used in the organization of work. In other words, as required by a risk assessment in circumstances where risks to health and safety have not been avoided by other means, for example engineering controls or safe systems of work. Signs must be standardized across Europe in a way in which will reduce the hazards which may arise from linguistic and cultural differences between workers.

Safety signs are to warn of any remaining significant risk or to instruct employees of the measures they must take in relation to these risks. It is very important that employees fully understand the meaning of such safety signs and are aware of the consequences of not following the warning or instruction given by the sign.

It should be noted that arc flash labelling and field marking must conform to Directive 92/58/EEC and in fact labels based on the U.S. ANSI Z535 Standard would not be acceptable in most instances.

16.3 UK and Ireland

Most legislation in member states is driven by acts and regulations backed up by codes of practice and voluntary guidance based on national and international standards. It is through this local legislation that the EU Framework Directive is translated.

The cornerstone of this legislation is risk assessment and there is a move away from prescriptive measures which may involve a direct link from the arc flash hazard into PPE. In other words, the goal is to eliminate, minimise or mitigate the hazard before considering PPE. A result of this is a much lower tolerance of live working for workers in the industrial sectors and even utility workers will have to justify working on or near energized conductors. For instance, in Ireland and the United Kingdom the regulations covering live working are stated as follows.

Regulation 14 of the UK Electricity at Work Regulations 1989 (EAWR) and 86d of the Irish Safety Health and Welfare at Work (General Application) Regulations requires that three conditions are met for live working to be permitted where danger may arise. It is stressed that if just one of these conditions cannot be met, live working cannot be permitted, and dead working is necessary. The conditions are:

a) It is unreasonable in all the circumstances for the conductor to be dead; and
b) It is reasonable in all the circumstances for the person to be at work on or near that conductor while it is live; and
c) Suitable precautions (including, where necessary, the provision of personal protective equipment) have been taken to prevent injury.

This rigorous test of reasonableness even refers to diagnostic testing of live conductors.

Contrast this against the requirements of US legislation expressed in OSHA 1910.333 (a) which states that the employer cannot allow work to proceed "energized" or "live" except when it can be demonstrated that:

a) De-energizing introduces additional or increased hazards.
b) Infeasible due to equipment design or operational limitations.

16.3.1 UK Corporate Manslaughter Act

The Corporate Manslaughter and Corporate Homicide Act (2007) introduces a new offence for prosecuting companies and other organizations where there has been a gross failing throughout the organization in the management of health and safety with fatal consequences. Although not a European Directive and only enforceable in the UK, the penalties for those convicted under the new legislation are extremely severe. The fines have no upper limit and there is also a publicity order which means that the company will have to publicise the details of the conviction. The former damages the company balance sheet but the latter damages the company brand which, in many cases is PRICELESS.

The new offence came into force in 2008 and complements the existing law where individuals can be prosecuted for gross negligence manslaughter/culpable homicide and for health and safety offences. The Act does not change this and prosecutions against individuals will continue to be taken where there is sufficient evidence and it is in the public interest to do so. The new offence applies to all companies and other corporate bodies operating in the UK, whether incorporated in the UK or abroad.

16.4 Examples of specific legislative requirements satisfied by arc flash analysis

The following are some examples of where an arc flash hazard analysis as part of a risk assessment can help to comply with the above legislation.

1. Satisfy legal requirements for risk assessment which is embodied in law through Directive 89/391 (EU Workplace Health and Safety Directive) and how this directive is interpreted in secondary legislation in member countries such as Regulation 3 of the UK Management of Health and Safety at Work Regulations 1999. Analysing the risks requires some measure of severity of the arc flash hazard. In both the Directive and the Regulations mentioned, an evaluation of risks that cannot be avoided is required by the Principles of Prevention.

2. Any interactions which may be perceived as live work, including diagnostic testing need to be justified in Great Britain and Northern Ireland separately under regulation 14 of the Electricity at work regulations. Safety Health and Welfare at Work (General Application) Regulations 2007 Part 3 – Electricity apply in the Republic of Ireland.

3. By doing the hazard severity calculations, the onset of a second-degree burn, or partial thickness burn can be predicted at a distance from the possible arcing source. The distance at which the onset of a second-degree burn is calculated is called the arc flash boundary and is used within US and Canadian safety standards to determine the measures that will be required to protect workers. In the UK, Designers may find that this information will help with a quantifiable approach to working space requirements of regulation 15 of the Electricity at Work Regulations 1989. (NI 1991) and 87 of the Irish Safety Health and Welfare at Work (General Application) Regulations.

4. The determination of prospective short circuit current levels is required by regulation four and five of the Electricity at Work Regulations 1989. (NI 1991)

5. Technical knowledge about the system to be worked on or near is a key ingredient in the competence of the individual undertaking that work. HSR25 Memorandum of guidance on the Electricity at Work Regulations 1989 states "The scope of technical knowledge or experience may include:
   a) adequate knowledge of electricity;
   b) adequate experience of electrical work;
   c) adequate understanding of the system to be worked on and practical experience of that class of system;
   d) understanding of the hazards which may arise during the work and the precautions that need to be taken;
   e) ability to recognise at all times whether it is safe for work to continue."

All this provides good information about how severe the outcome of an unplanned event will be. Although the aim must always be to avoid the hazard in the first place, there is now the technology available, based on sound research, to determine the protection measures for workers.

6. To be able to comply with the Electricity at Work Regulations 4 and 14 and 86 (1d) and (2a) of the Irish Safety Health and Welfare at Work (General Application) Regulations, the selection of control measures including PPE would be very difficult to achieve without a measure of the actual hazard.

7. Ensure compliance with European Council PPE Directive 89/656/EEC and UK Personal Protective Equipment at Work Regulations 1992 (NI 1993) in carrying out a risk assessment. "PPE can only be prescribed after the employer has analysed the risks which cannot be avoided by other means." In addition, by following the 4P steps in this guide, this will help to ensure that measures can be taken to reduce incident energy levels and therefore remove some of the issues to do with sensory deprivation. Before choosing personal protective equipment, Article 5 of 89/656/EEC asks us to assess whether the PPE that we intend to use will satisfy the following requirements.

   a) be appropriate for the risks involved, without itself leading to any increased risk;

   b) correspond to existing conditions at the workplace;

   c) take account of ergonomic requirements and the worker's state of health;

   d) fit the wearer correctly after any necessary adjustment.

16.5 Brexit

The UK has left the European Union at the time of writing and the following is my understanding of how this will affect electrical safety regulations and law. Regardless of final trading relationships, the HSE have provided guidance that the ultimate aims of protecting people at work will not change regardless of the outcome. The main role of the EU health and safety legislation has been to harmonise legal standards and to remove barriers to trade among member nations. Within the UK and EU, health and safety legislation is based upon risk assessment which is unlikely to change in the near future. None of the UK legislation listed above will change but clearly the EU directives will not apply directly. However, to ensure the continued stability of the European standardization system and to give confidence to European business and consumers, CEN and CENELEC Members have agreed, to extend the current transition period for their UK Member, BSI, until 31 December 2021.

In fact, very few of the existing health and safety regulations are to be amended as a direct result of Brexit. I have however, created a table, Figure 16.1 below, which may help to identify electrical safety related legislation and equivalents within the European Directives. To read more about amended regulations, please visit the HSE Website.

https://www.hse.gov.uk/brexit/

The primary legislation in the Great Britain continues to be the Health and Safety at Work etc. Act 1974 and similar legislation in Northern Ireland, the Health and Safety at Work (Northern Ireland) Order 1978. The Health and Safety at Work etc Act 1974 is an enabling act which means that secondary legislation or regulations can be passed such as the Electricity at Work Regulations 1989. It sets out the general duties which: employers have towards employees and members of the public; employees have to themselves and to each other and certain self-employed individuals have towards themselves and others.

Figure 16.1

Similar Legislation across the EU, Ireland and UK

European Council Directive	Republic of Ireland Legislation	Great Britain and Northern Ireland Legislation
89/391/EEC Workplace Health and Safety Directive (Framework Directive).	Safety, Health and Welfare at Work Act 2005.	Management of Health and Safety at Work Regulations 1999. Management of Health and Safety at Work Regulations (Northern Ireland) 2000.
Electricity at Work - No Specific Directive.	Safety Health and Welfare at Work (General Application) Regulations 2007 Part 3 - Electricity.	Electricity at Work Regulations 1989. Electricity at Work Regulations (Northern Ireland) 1991.
EU Council Directive 1989/656/EEC Use of Personal Protective Equipment.	Safety Health and Welfare at Work (General Application) Regulations 2007 Part 2 WORKPLACE AND WORK EQUIPMENT Chapter 3—Personal Protective Equipment.	Personal Protective Equipment at Work Regulations 1992. Personal Protective Equipment at Work Regulations (Northern Ireland) 1993.
EU Council Directive 2009/104/EC Use of Work Equipment.	Safety Health and Welfare at Work (General Application) Regulations 2007 PART 2 WORKPLACE AND WORK EQUIPMENT Chapter 2— Use of Work Equipment.	Provision and Use of Work Equipment Regulations (Northern Ireland) 1999). Provision and Use of Work Equipment Regulations 1998.
EU Council Directive 1992/58/EEC Safety and/or Health Signs.	Safety Health and Welfare at Work (General Application) Regulations 2007 PART 7 SAFETY SIGNS AND FIRST AID and Chapter 1— Safety Signs at Places of Work.	Health and Safety (Safety Signs and Signals) Regulations 1996. Health and Safety (Safety Signs and Signals) Regulations (Northern Ireland) 1996.

16.6 The International Electrotechnical Commission (IEC)

The International Electrotechnical Commission (IEC) is the world's leading organization that prepares and publishes International Standards for all electrical, electronic and related technologies. Individuals like myself participate in the IEC's work through National Committees (NCs) and there can only be one per country. Many of the standards that are listed as IEC in the following table. They are extremely comprehensive and are the result of years of work from experts. IEC cooperates with CENELEC which is the European Committee for Electrotechnical Standardization and is responsible for standardization in the electrotechnical engineering field. CENELEC prepares voluntary standards, which help facilitate trade between countries, create new markets, cut compliance costs and support the development of a single European market.

16.7 Directory of Useful Legislation and Standards

The following table gives various standards that may assist with various parts of this guide. This includes standards that are not quoted directly in the guide but may assist the reader in locating information in order to apply some of the principles. For instance, the data collection section relates to electrical relays, current transformers, cables, busbars, switchgear etc, all of which have standards listed below in Figure 16.2.

Figure 16.2

Topic	Legislation or Standard	Description
Directory of Useful Legislation or Standards Electrical Safety or Arc Flash Related		
Health and Safety		
	EU Directive 89/391/EEC (Workplace Health and Safety Directive).	Often referred to as the "Framework Directive" introduced measures to encourage improvements to the safety and health of workers. A cornerstone of this directive is risk assessment which means that there are no prescriptive standards covering electrical safety, especially arc flash, and certainly no link from the hazard directly into arc flash PPE.
	EU Directive 2009/104/EC The Use of Work Equipment.	The Use of Work Equipment Directive 2009/104/EC defines the minimum health & safety requirements for the operation of work equipment by employees at work.
	EU Directive 92/58/EEC - Safety and/or health signs.	Council Directive 92/58/EEC defines the minimum requirements for the provision of health and/or safety signs at work and complements the Framework Directive 89/391/EEC on health and safety at work.
	INTERNATIONAL STANDARD ISO 31000 Risk management principles and guidelines.	Describes a generic approach to providing the principles and guidelines for managing any form of risk in a systematic, transparent, and credible manner and within any scope and context.

Topic	Legislation or Standard	Description
PPE		
	EU Directive 89/656/ EEC - use of personal protective equipment.	The PPE Directive sets the minimum requirements for the assessment, selection, and correct use of personal protective equipment.
	EU Regulation (EU) 2016/425 on personal protective equipment repealing Council Directive 89/686/EEC.	Design, manufacture, and marketing of personal protective equipment. This regulation ensures common standards for personal protective equipment (PPE) in all member States in terms of protection of health and the safety of users, while enabling the free movement of PPE within the Union.
	IEC 61482-2: Live working - Protective clothing against the thermal hazards of an electric arc - Part 2: Requirements.	Applicable to protective clothing used in work where there is the risk of exposure to an electric arc hazard. This document specifies requirements and test methods applicable to materials and garments for protective clothing for electrical workers against the thermal hazards of an electric arc.
	IEC 61482-1-1: Live working – Protective clothing against the thermal hazards of an electric arc. (Open arc testing)	Part 1-1: Test methods - Method 1: Determination of the arc rating (ELIM, ATPV and/or EBT) of clothing materials and of protective clothing using an open arc.
	IEC 61482-1-2:2: Live working - Protective clothing against the thermal hazards of an electric arc. (Box testing)	Part 1-2: Test methods - Method 2: Determination of arc protection class of material and clothing by using a constrained and directed arc (box test).
	IEC 60903 Electrical Insulating Gloves.	Live Working - Electrical Insulating Gloves. Is applicable to electrical insulating gloves and mitts that provide protection of the worker against electric shock.
	IEC 63247-1: Live working - Footwear for electrical protection - Part 1: Insulating footwear and over boots.	Specifies the requirements and testing for PPE footwear used as electrical insulating footwear and over boots that provide protection of the worker against electric shock and used for working live or close to live parts.
	IEC 62819 Live Working Eye Face and Head Protectors against the effects of an electrical arc, STANDARD AT DRAFT STAGE.	Eye Face and Head Protectors against the effects of an electrical arc, test and performance standards.

Topic	Legislation or Standard	Description
Live Working Equipment		
	IEC 60900: Live working - Hand tools for use up to 1000 V AC and 1500 V DC.	Is applicable to insulated, insulating and hybrid hand tools used for working live or close to live parts at nominal voltages up to 1 000 V AC and 1 500 V DC.
	IEC 61111:2009 Live working - Electrical insulating matting.	Is applicable to electrical insulating matting made of elastomer for use as a floor covering for the electrical protection of workers on electrical installations.
	IEC 61010 - Safety requirements for electrical equipment for measurement, control, and laboratory use - Part 1: General requirements.	Specifies general safety requirements for the following types of electrical equipment and their accessories, wherever they are intended to be used. a) Electrical test and measurement equipment. b) Electrical industrial process-control equipment. c) Electrical laboratory equipment.
	IEC 61112:2009 Live working - Electrical insulating blankets.	Applicable to electrical insulating blankets for the protection of workers from accidental contact with live or earthed electrical conductors, apparatus or circuits and avoidance of short circuits on electrical installations.
System Studies		
	IEC 60909-0 Short-circuit currents in three-phase a c. systems.	The calculation of short circuit currents in low voltage three-phase AC systems, and in high-voltage three-phase AC systems, operating at a nominal frequency of 50 Hz or 60 Hz. Method mostly used in Europe.
	IEEE 1584 Guide for Performing Arc Flash Hazard Calculations.	IEEE 1584 is widely seen within the worldwide electrical engineering community as a valid means of determining incident energy levels and arc flash boundaries and calculations are auditable against this standard. The standard does not offer any advice or recommendations in respect of personal protective equipment or any other mitigation measures.
	DGUV -I- 203-5188 E Guide to the Selection of Personal Protective Equipment for Electrical Work.	Often referred to as a European/German Standard, the calculations used by DGUV-I 203-078 are derived from the box test method for PPE which is detailed in the standard IEC 61482-1-2:2014 Live Working - Protective Clothing against the thermal hazards of an electric arc.
	IEC 60529 Ingress Protection Code.	The IP Code, or Ingress Protection Code, classifies and rates the degree of protection provided by mechanical casings and electrical enclosures against intrusion, dust, accidental contact, and water.

Topic	Legislation or Standard	Description
Switchgear and Control Gear		
	BS 6626 (Maintenance of switchgear from 1 kV to 36 kV).	British Standard which gives recommendations and guidance for the maintenance of electrical switchgear and control gear having a rated AC or DC voltage above 1 kV and up to and including 36 kV
	IEC 60898 Electrical accessories – Circuit-breakers for overcurrent protection for household and similar installations.	Miniature circuit breakers (MCB) for overcurrent protection of residential and similar installations. Rated currents range from 0.16 A to 125 A. The short circuit breaking capacity is up to 25 kA.
	IEC 60947 Series (LV Switchgear and control gear).	The international standard applies, when required by the relevant product standard, to low-voltage switchgear and control gear not exceeding 1000 V AC or 1500 V DC.
	IEC 61439-1 (LV switchgear Assemblies: General).	General Rules for low voltage switchgear and control gear assemblies.
	IEC 61439-2 (LV switchgear: Power switchgear and control assemblies).	Low voltage switchgear and control gear assemblies up to 1000 volts AC and 1500 volts DC. Up to 1000 Hz.
	IEC 61439-6 (LV Bus Bar Trunking).	Low voltage bus bar trunking.
	IEC 62271-1 (HV switchgear: Common Clauses).	High voltage switchgear and control gear - Part 1: Common specifications for alternating current switchgear and control gear from 1kV to 52kV.
	IEC 62271-100 (HV switchgear: Circuit- breakers).	High-voltage switchgear and control gear - AC circuit breakers.
	IEC 62271-102 (HV switchgear: Disconnectors and earth switches).	High voltage alternating current disconnectors and earthing switches.
	IEC 62271-103 (HV switchgear: Switches).	Switches for rated voltages above 1 kV up to and including 52 kV.
	IEC 62271-200 (HV switchgear).	AC metal-enclosed switchgear and control gear for rated voltages above 1 kV and up to and including 52 kV.
	IEC 60076 Series (Power transformers).	Power Transformers and reactors of all types and testing.
	IEC 60269 (all parts) Low voltage fuses.	The standard unifies several national standards, such as DIN and BS88. All fuses of different technologies tested to meet IEC standards will have similar time-current characteristics, which simplifies design and maintenance.

Topic	Legislation or Standard	Description
Protection		
	IEC 60282 Series (HV fuses).	High-voltage fuses, including Current-limiting fuses and expulsion type fuses.
	IEC 61869 Series (Instrument transformers).	Requirements for instrument transformers including current transformers.
	IEC 60255 Series (Measuring relays and protection equipment).	Measuring relays and protection equipment, including sections for types, testing and electromagnetic compatibility.
Cables		
	IEC 60287 Series.	Calculation of current ratings for cables.
	BS 7870 Series (LV and MV polymeric insulated cables for se by utilities).	British Standard BS7870 covers low voltage and medium voltage cables used for the provision of an electricity supply from a distribution main.
	IEC 60502 Series (Power cables from 1kV to 30kV).	Power cables with extruded insulation and their accessories for rated voltages from 1kV to 30kV.
Overhead Lines		
	EN 50341-1 (Overhead lines exceeding 1kVAC).	Overhead electrical lines exceeding AC 1 kV - Part 1: General requirements - Common specifications General This European Standard applies to new overhead electric lines with nominal system voltages exceeding AC 1 kV and with rated frequencies below 100 Hz.
Arc suppression and detection		
	IEC 60947-9-1: Low-voltage switchgear and control gear - Part 9-1: Active arc-fault mitigation systems - Arc quenching devices	Covers low voltage arc quenching devices, hereinafter referred to as AQDs, which are intended to eliminate arc faults in low voltage assemblies (typically low-voltage switchgear and control gear assemblies in accordance with the IEC 61439 series), by creating a lower impedance current path, to cause the arcing current to transfer to the new current path.
	IEC 62606 General requirements for arc fault detection devices.	Applies to arc fault detection devices (AFDD) for household and similar uses in AC circuits.
	IEC TS 63107: Integration of internal arc-fault mitigation systems in power switchgear and control gear assemblies.	IEC TS 63107 states the requirements for integration and testing of internal arc fault mitigation systems in low voltage switchgear and control gear assemblies – PSC-assemblies, according to IEC 61439-2 to demonstrate their correct operation.

Topic	Legislation or Standard	Description
Electrical Safety		
	EN 50110-1, Operation of Electrical Systems.	This European Standard is applicable to all operation of and work activity on, with, or near electrical installations. These are electrical installations operating at voltage levels from and including extra-low voltage up to and including high voltage.
	NFPA 70E Standard for Electrical Safety in the Workplace.	The National Fire Prevention Association of America's standard for Electrical Safety in the Workplace. NFPA 70E requirements for safe work practices to protect personnel by reducing exposure to major electrical hazards.

16.8 DGUV -I- 203-5188 E Guide to the Selection of Personal Protective Equipment

The DGUV-I 203-5188 E Guide to the Selection of Personal Protective Equipment for Electrical Work has been developed and used in Germany for the section of Arc Rated Personal Protective Equipment. DGUV is the German Social Accident Insurance Institution which is the national, compulsory program that insures workers for injuries or illness incurred through their employment.

At the core of the guide, is an algorithm that has been derived from the standard IEC 61482-1-2:2014, Method 2: Determination of arc protection class of material and clothing by using a constrained and directed arc (box test). This testing method for PPE is described in Chapter 7: Protection. The information provides guidance for the appraisal of potential thermal hazards due to electric fault arcs associated with electrotechnical work on electrical equipment. Accordingly, the document affords employers an element of support in the selection of essential PPE and is applicable for work performed on or in the vicinity of electrical equipment greater than 50 V AC. (Note that there is a new version of the guide that has been published but in German only. This gives an upper limit of 110kV and also has a section on DC calculations up to 1500 volts)

Whilst the document is describing the algorithm as a European method, this is not the case for industrial and commercial installations where the IEEE 1584 Guide for Performing Arc Flash Hazard Calculations is more appropriate.

*BGI/GUV-I 5188 E available from the German Social Accident Insurance Institution or on
https://publikationen.dguv.de/dguv/udt_dguv_main.aspx

16.9 NFPA 70E Standard for Electrical Safety in the Workplace

*NFPA 70E requirements for safe work practices to protect personnel by reducing exposure to major electrical hazards. As described by the National Fire Prevention Association of America, the scope of the standard addresses electrical safety-related work practices, safety-related maintenance requirements and other administrative controls for employee workplaces that are necessary for the practical safeguarding of employees relative to the hazards associated with electrical energy during activities such as the installation, inspection, operation, maintenance, and demolition of electric conductors, electric equipment, signalling and communications conductors and equipment and raceways. NFPA 70 E was created in 1979 at the request of the US Occupational Safety and Health Administration and as a result, even though it is a consensus standard, can be said to have a quasi-legal standing in the United States of America.

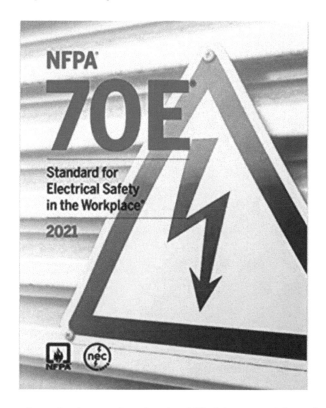

The standard is formally reviewed every three years and I have seen how it has evolved from hazard protection to risk prevention. In fact, it has at its heart, elimination as a first principle. It can appear to be a big document at first glance but actually the main chapter (1) on safety related work practices is only 19 pages long. This covers all the general requirements, establishing an electrically safe work condition and work involving electrical hazards including arc flash. The rest of the document includes maintenance, special equipment and 17 informative annexes.

There are a lot of positive attributes but there has got to be some caution and explanation when mentioning the standard. In the UK we do not have standards for rules (except at high voltage) and some of the NFPA 70E requirements may contradict HSE guidance. For example, NFPA 70E has requirements for live working electrical permits. In the UK this may risk an improvement notice as electrical permits must never be used for live work. Requirements for warning labels will be at odds with UK norms through the signs and signals legislation. That said, where I have had US owned clients in Europe, I have been successful in steering a course between a head office requirement to impose NFPA 70E into overseas operations and the local regulatory requirements.

*The NFPA 70E Standard for Electrical Safety in the Workplace is available from
https://www.nfpa.org/codes-and-standards/all-codes-and-standards/list-of-codes-and-standards/detail?code=70E

Bibliography

Please see also Chapter 16 for Rules Codes and Legislation that are not referenced directly in the guide but are relevant to arc flash hazard measurement and management.

1. Introduction to Health and Safety at Work by Phil Hughes and Ed Ferrett (NEBOSH National General Certificate textbook).
2. European View of Arc Flash Hazards and Electrical Safety by Mike Frain and Jim Phillips paper presented to the IEEE Electrical Safety Workshop 2012 Daytona Beach Florida.
3. National Fire Protection Association Standard for Electrical Safety in the Workplace NFPA70E.
4. IEEE 1584 Guide for Performing Arc Flash Calculations. The Institute of Electrical and Electronics Engineers.
5. European Council Directive 89/391/EEC. (EU Workplace Health and Safety Directive)
6. European Council Directive 89/656/EEC (Use of Personal Protective Equipment)
7. The Other Electrical Hazard: Electrical Arc Blast Burns– Ralph H Lee IEEE Transactions 1982.
8. Pressures developed by Arcs – Ralph H Lee IEEE Transactions 1987.
9. German guide DGUV-I 203-078 Thermal Hazards from Electric Fault Arc
10. IEC 61482-1-2:2014, Live working - Protective clothing against the thermal hazards of an electric arc - Part 1-2: Test methods - Method 2: Determination of arc protection class of material and clothing by using a constrained and directed arc (box test)
11. IEC 61482-1-1:2019, Live working – Protective clothing against the thermal hazards of an electric arc – Part 1-1: Test methods – Method 1: Determination of the arc rating (ELIM, ATPV and/or EBT) of clothing materials and of protective clothing using an open arc.
12. IEC 60269-1 Low voltage fuses
13. IEC 61439-2 Power switchgear and control gear assemblies (PSC)
14. Considerations for Using High-Impedance or Low-Impedance Relays for Bus Differential Protection -Ken Behrendt, David Costello, and Stanley E. Zocholl
15. GAMBICA Technical Guide low-voltage power switchgear and control gear assemblies Edition 2: 2013 A guide to the requirements of: BS EN 61439-1:2011 BS EN 61439-2:2011
16. Technical report IEC/TR 61641 - Enclosed low-voltage switchgear and control gear assemblies – Guide for testing under conditions of arcing due to internal fault.
17. A GAMBICA Technical guide internal arc fault capability of assemblies in accordance with IEC/TR 61641
18. Regulation (EU) 2016/425 of the European Parliament and of the Council of 9 March 2016 on Personal Protective Equipment
19. European Council Directive 89/656/EEC Use of Personal Protective Equipment
20. ASTM F2178 / F2178M - 20 Standard Specification for Arc Rated Eye or Face Protective Products.

21. The BEAMA (British Electrotechnical and Allied Manufacturers' Association) Guide to Forms of Separation.
22. Forms of Separation - a time for change - Schneider Electric 2017
23. IEC 62271-200 High-voltage switchgear and control gear
24. The Health and Safety (Safety Signs and Signals) Regulations 1996
25. European Directive 92/58/EEC Safety and/or Health Signs
26. The Republic of Ireland Guide to the Safety, Health and Welfare at Work (General Application) Regulations 2007 (Part 3 Electricity)
27. Guideline for Assessing the Competence of Persons involved in Live Working published by the International Social Security Association (ISBN-Nr. 3-9807576-6-8).
28. Guidance Note GS38 (Fourth edition) on electrical test equipment for use on low voltage electrical systems.
29. Code of Practice for Electrical Safety Management – Institution of Engineering and Technology ISBN 9781849196697
30. EN 50110-1:2013 Operation of Electrical Installations.
31. Observations on the procedures for testing the protective effect of Personal Safety equipment against the effect of arcing – Dr Dirk Borneburg, Dr Frank Wachholz, Jurgen Volger and Dr Helmut Eichinger 2007.
32. Arc Flash Forum on www.arcflashforum.com.
33. Mitigating Personnel Exposure to The Arc Flash Hazard. White paper written by Catherine Irwin, Udo Kerssebaum, Helmut Eichinger, Daniel Doan June 2009.
34. Fear of Flashover - article written for the Institute of Engineering and Technology Power Engineer magazine by Mike Frain and Jim Phillips July 2007.
35. Electrical Flashover-Live Proximity Work in Industry-Magazine article written for the Electrical review by Mike Frain, September 2007.
36. Arc Flash-not just electrocution hazard, parts 1,2 &3 – A trilogy of magazine articles by Mike Frain and Jim Phillips written for the Electrical Review September, October & November 2009.
37. Guidance for Assessing the Competence of Persons involved in Live Working Issue by the International Social Security Association.
38. HSR25 Memorandum of Guidance on the Electricity at Work Regulations 1989 (HSE books www.hse.gov.uk)
39. HSG85 Electricity at Work - Safe Working Practices (HSE books www.hse.gov.uk)
40. HSG230 Keeping Electrical switchgear Safe (HSE books www.hse.gov.uk)
41. INDG372 Electrical Switchgear and Safety (HSE books www.hse.gov.uk)
42. INDG163REV2 Five Steps to Risk Assessment (HSE books www.hse.gov.uk)
43. INDG 354 Safety in Electrical Testing at Work (HSE books www.hse.gov.uk)
44. GS38 Electrical Test Equipment for use by Electricians (HSE books www.hse.gov.uk)
45. Comprehensive Overview and Comparison of ANSI vs. IEC Short Circuit Calculations: Using IEC Short Circuit Results in IEEE 1584 Arc Flash Calculations". A. Majd, R. Lou, M. Devadass, J. Phillips, IEEE PCIC Conference proceedings, Cincinnati, OH Sept. 2018.
46. Understanding Short Circuit Motor Contribution by Darrell Broussard, GE Industrial Solutions.

47. Complete Guide to Arc Flash Hazard Calculation Studies by Jim Phillips P.E – Brainfiller Inc. 2010.
48. DC Arc Flash Calculations by Jim Phillips – Posted on Brainfiller.com, November 20, 2019.
49. Arc Flash Calculations for Exposures to DC Systems by D.R. Doan – IEEE Transactions on Industry Applications, Vol.46 No 5.
50. DC Arc Models and Incident Energy Calculations by R.F. Ammerman, T. Gammon, P.K. Sen and J.P. Nelson – IEEE Transactions on Industry Applications, Vol.46 No 6.
51. Electric Arcs in open Air by A.D.Stokes and W.T.Oppenlander – Journal of Physics D: Applied Physics 1991.
52. Simple Improved Equations for Arc Flash Analysis by R. Wilkins – IEEE Electrical Safety Forum, August 30, 2004.
53. Schneider Electric Cahier technique no. 158 - Calculation of short-circuit currents by Metz-Noblat, Dumas and Poulain.
54. Method and Rating System for the Evaluation of Thermal Protection, Stoll,A.M. and M.A. Chianta

Printed in the USA
CPSIA information can be obtained
at www.ICGtesting.com
LVHW071316150324
774517LV00056B/2668